# PHILO MARSHALL EVERETT

*Father of*

*Michigan's Iron Industry*

*and*

*Founder of the*

*City of Marquette*

by

*His Great-Grandson*

# FRANK B. STONE

GATEWAY PRESS, INC.
Baltimore, MD    1997

*Frontispiece: A Portrait of Philo Marshall Everett,*
*a photograph by Childs, about 1872.*

Please direct all correspondence to:
Frank B. Stone
24 Ramsey Drive
Summit, New Jersey 07901-3015

To be sold and distributed by:
SNOWBOUND BOOKS
118 N. Third Street
Marquette, MI 49855

Library of Congress Catalog Card Number 96-79722
ISBN 0-9655955-0-1

Stone, Frank B. (1913 -     )
1.  Everett, Philo Marshall (1807-1892)
2.  Michigan - Marquette History (1840-1892)
3.  Iron Mining - Michigan
4.  Lake Superior - Exploration
5.  Jackson Mine

Published for the author by
Gateway Press, Inc.
1001 N. Calvert Street
Baltimore, MD 21202

*Printed in the United States of America*

# CONTENTS

The Biography

# LIST OF ILLUSTRATIONS

The maps (Plates 1, 2, 11, 12, 13, 14, 17, 18, 19, 21, 22, 23, 27 and 33) are from the author's collection. The map used in Plate 11 and the work referred to are from the Library of Michigan Technological University (herein MTU). The map used in Plates 17, 18 and 27 was given him many years ago by Mrs. Charlotte Dively, a grand-daughter of J.M.Longyear. An original is in the Longyear Research Library at the Marquette County Historical Society (herein MCHS).

Various photographs (originals of which are not owned by the author) were obtained as follows:

Plates 7, 16, 24, 25 and 29 were reproduced by Jack Deo, owner of Superior View, 137 1/2 West Washington Street, Marquette, MI from his very large collection of historical photographs. Plates 28 and 37 are pictures taken by him specially for this volume.

Plates 20,31 and 35 are from the collection of the
Marquette County Historical Society (herein MCHS).

.Plates 5, 30 and 34 are from family papers deposited
with the Michigan Iron Industry Museum (herein MIIM).Some
of them have also been deposited with MCHS.

See ACKNOWLEDGMENTS (below) for a further description
of the help received from these sources and others.

# FOREWORD

"On arriving at Teal Lake, we found the ore.
... There lay the boulders of the trail,
made smooth by the atmosphere, bright and
shining, but dark colored, and a
perpendicular bluff, fifty feet in height,
of pure solid ore..."

Thus did Philo Marshall Everett describe his
first discovery of the bounty of iron ore in the
wilds of Northern Michigan; a bounty which he
believed would supply all the iron the world would
ever need.

You are about to read a story of the birth of
the iron ore mining industry in Michigan's majestic
Upper Peninsula. It is the story of Philo Marshall
Everett, written by his great grandson, and his
dramatic efforts to unlock the deposits of iron ore
buried on the isolated shore of Lake Superior in the
1840's. Using extensive excerpts from his great
grandfather's letters and writings, as well as those
from other family members, Frank B. Stone has brought
this remarkable pioneer and his era vividly to life.

Mr. Stone has provided a wealth of maps, sketches and photographs -- many of them rare, some heretofore unknown -- to illustrate this fascinating history. Here unfolds a vivid picture of the severe hardships and challenges which confronted Mr. Everett as he struggled to mine the ore and transport it from a wilderness devoid of roads, railroads or shipping ports. What also emerges is an intimate glimpse of the natural world around him at that time.

Beginning with Everett's work as a young bargeman and engineer on the Erie Canal, the reader is able to share Everett's impressions of the fur-trading markets in Sault Ste. Marie; the arduous journey from Detroit to Mackinac by steamer and by small boat on to Lake Superior; the development of the first iron foundry; the excruciating problems of transporting the ore, first by horse and then by mule teams over a hand-built but often impassable plank road; the six-month wait for spring when ships could finally bring in sorely needed supplies and mail. This history records Everett's interaction with other early Marquette pioneers and residents like William Austin Burt, surveyor and inventor of the solar compass, and his son John Burt, who remained as a developer of Marquette; Louis Nolan, Everett's French-Canadian guide, coaster and expert on Lake Superior; and Marji-Gesick, the Chippewa Indian chief who led Everett to "the mountain of iron".

The story of Philo Marshall Everett is also the story of the early history of Marquette County, which grew from a small settlement into a Great Lakes shipping port, railroad center and mining capital and which would eventually produce most of the iron used during the great Civil War. It chronicles the emergence of the banking, railroad and commerce industries in Marquette; the development of its churches, schools and courts as well as the devastating effects of the Great Fire of 1868, which razed the downtown buildings and shipping docks of Marquette. Also the descriptions of early Marquette social and family life by Mehitable Everett are particularly insightful.

Here, too, are impressive descriptions of Lake Superior. As a sailor, I especially appreciated the descriptions of the "Big Lake", with its daunting coastline and its treacherous storms. One can almost see the schooners and brigantines under full sail approaching the shelter of Marquette's harbor and hear the welcoming shouts and cheers of the people on shore.

Author Stone has produced an inspiring tale of the adventurous founding of a vitally important industry and region; and of the legacy of one man's courage, resourcefulness and perseverance. It is a

tribute to the resiliency of the human spirit in the
face of almost unimaginable adversity.


Those of us fortunate enough to live in
Marquette and others who seek a depth of knowledge
and appreciation for this majestic area owe Frank
Stone a huge debt for this endeavor. As Mehitable
Everett, Philo's wife, observed in 1879: "It often
brightens the present to look back on the past.

PATRICIA L. MICKLOW

Marquette, Michigan
May 6, 1996

# ACKNOWLEDGEMENTS

A reading of this biography and its footnotes will show that it has been assembled like a jigsaw puzzle from many records, mostly from the last century. The footnotes may also serve as a bibliography, sometimes suggesting the comparative value of the respective sources.

It would be too long a job to relate the steps by which the work was put together and to give adequate thanks to the many people who helped. Hence, this may seem like a mere list.

My mother, Grace Ball Stone, started me with many affectionate stories about her grandfather, ultimately documented in her published and unpublished memorabilia, e.g. the Everett Recollections, C.R.Everett's articles, pictures, etc.

In correspondence I learned about Ernest Rankin's interest and the material that he had collected from the family and others for the Marquette County Historical

Society (MCHS), particularly about a study made by Patricia
L Micklow when a student at Northern Michigan University
(NMU). This was the most accurate and understanding work on
Philo Marshall Everett that anyone had done, not even
excepting the works of family members. Now Judge Micklow,
she has continued her interest and has been most supportive
of my work, as appears in the FOREWORD that she has
contributed. Others at NMU have also been helpful,
particularly Russell Magnaghi and James Carter.

At the MCHS and its J.M.Longyear Research Library
I found extensive files on the Everett, Ball and other
families, together with much, much more on the history of
Marquette, such as newspapers, letters and photographs. Here
John M. Maitland, a former director, and Linda K. Panian,
Librarian, were very helpful. Ms. Panian did substantial
research and referred me to Catherine Grady, who worked on
land records. Mrs. Charlotte M. Dively was also encouraging
and urged me on.

While a number of people had studied the judicial
decisions in the Kawbawgam litigation, the Transcript used
in one of those appeals proved to be the bedrock story of
the discovery of the "iron mountain" by Everett and his
party. The Transcript, a copy of which was obtained via
MCHS, was  a source that seemed not to have been used
before.

As noted within, Frank Matthews showed me the
contents of his Jackson Mine Museum. Then David Bridgens
supplied the Report on the Carp River Forge that promoted
the Michigan Iron Industry Museum (MIIM). There the history

of the industry and Everett's participation in it were for the first time fully displayed. Not only that, but Thomas Friggens made available the extensive family papers that C. R. Everett and my mother had preserved and which were deposited there by my second cousin, Marion Everett Cole. The unpublished memoir that my mother had supplied was a new and vital source.

At the Van Pelt Library of Michigan Technological University (MTU), I had the help of Archaeologist Patrick Martin, Librarian Teresa Spence and Archivist Erik C. Nordberg. At Pat and Peter Van Pelt's Museum Shop in Eagle Harbor, I found many books, especially the Gray Report described within, which more fully explained my precious map shown in Plates 2, 12 and 13.

For nearly two years William Trevarrow, recently Chairman of the Publications Committee of MCHS, has supplied the most (and the most indispensable) help and support.

Frank B. Stone

24 Ramsey Drive
Summit, New Jersey
   07901-3015

  June 1996

## DEDICATION

Over a century has rolled by since the passing of Philo Marshall Everett. All of his children and grandchildren and most of his great grandchildren have followed him in death.

Now the author, the youngest of this remnant, has undertaken to reconstruct this biography. There are several reasons:

1. The role of "father" of the great Mid-Western Iron Industry, for which Everett has been named, is a reason in itself. He has been so recognized because he identified and staked out the claim to the first commercial body of iron ore to be found in all the area — the Jackson Mine. The subsequent struggles to develop that first mine are a secondary part of the story.

His contemporaries recognized how this discovery led to the development of the industry and of the Marquette area and so it seems unnecessary to argue about who was "first" in all these things. Nevertheless, such a competition has led to many conflicting stories over the last century. It is the purpose of this biography to cut through this maze -- to put aside the popularized fictions and obvious fantasies and to rely strictly upon established facts. Fortunately, there were judicial hearings and decisions from the Supreme Court of Michigan to form the bedrock of the true story.

2. Because he realized from the first that the bringing of the iron to market was a huge endeavor, Everett arranged to bring in another venturer with more resources than he and his twelve neighbors could supply. Even so, he and most of the others gave up the struggles of the mine itself and sold out at an early stage. At that point Everett (again with partners) claimed 36 acres on the shore of what became Iron Bay and developed the shipping port for iron that became the City of Marquette.

3. In this manner Everett became the real founder of Marquette, with the first home and then a larger house, where he held organization meetings for the County as the first Probate Judge and as the Chairman of the Township Supervisors. In these homes the Everetts welcomed other pioneers, such as the John Burt family, prospective investors from as far away as Boston, the Elys and Senator Sumner, as they planned the first railroad and, when he came to Marquette after his triumph with the "Soo" Canal, its builder, Charles Harvey. They continued to entertain their friend, the "Old Chief" Marji Gezick, the leader of the Chippewas in all the area.

Everett and his wife were also the organizers of St. Paul's Episcopal Church and its principal support at the outset. As Senior Warden for all his active life, he saw it grow into a cathedral under the able administration of his friend, Bishop Williams.

4. The full story of growth from a boy on the farm to this later status is found only in unpublished family papers. The author has called upon his lifelong immersion in American history to help to coordinate Everett's incomplete memoirs into the annals of the burgeoning economy of the early Republic. Everett experienced the great national expansion from the New York State Canals and other "internal improvements" to the first railroads in Michigan and, above all, to the establishing of a great industry, requiring the unlocking of Lake Superior and its mineral wealth. These momentous events are not much remembered in the nation at large. In the current consciousness the Upper Peninsula has become a step-child of Michigan, while its mineral wealth has been eclipsed by the industries of the "Rust Belt" that feed upon it.

The opening of the Old Northwest, so far as Lake Superior was concerned, had no Indian Wars to create drama. Even its mineral rushes, the most important in the nation, did not have the notoriety and violence that marked the Gold Rush of '49

2

and its ancillaries. In contrast to the saga of the Wild West and the supreme drama of the Civil War, there is no room on the television screen for the quieter history.

In 1973 and 1974 the Michigan History Division of the Department of State conducted a project at the site of the historic Jackson (Carp River) Forge (later described). They recognized that "the iron industry of the Lake Superior region began with the first bloom produced by the Jackson Forge" and that the development of the Lake Superior iron ranges made the dramatic increase in America's iron industry possible in the latter part of the Nineteenth Century. By that time, they pointed out "more than two-thirds of the nation's iron ore came from the Lake Superior Region". They added: "The greatness of our nation rests in part on that industrial base yet we know too little about our industrial history."

In their publication Carp River Forge:A Report, (herein "Forge",undated but issued shortly thereafter),they proposed as a 1976 Bicentennial project "the reconstruction of the first iron forge in Michigan and the creation of a museum of mining technology". Today the Michigan Iron Industry Museum in Negaunee is fulfilling that proposal.

The birth of the basic mining industry, which called for building the world's busiest canal, has indeed too long been a forgotten story. Everett's part in it was certainly not dominant. As an individual he merely showed how one man's effort could start the ball rolling. Everett led a full and rewarding life, even though at long last he was overcome by failing health and a nationwide panic. He was loved and revered by his fellows. As a recent writer[1] summed it up:

Philo Marshall Everett lived honestly and

in his quiet, innocuous way, founded a city

in the process.

# INTRODUCTION

This then is the story of a 19th Century American — one who lived in every decade of that century. Philo Marshall Everett was a pioneer in the Old Northwest (ie. the Northwest Territory, established by the Continental Congress before the United States had a Constitution).

We, who live at the end of the 20th Century, need some background explanations as we look back to 1807, the year Everett was born. His parents, like other colonists, had been confined for years to the boundaries of New England. Conflicts with the Indians and a Proclamation from King George held them there even after the French-supported wars were won. There was still the War of 1812 to be fought with Britain and its Indian allies before the Northwest Territory could be freed of massacres, made firmly American and fully opened up for settlement.

Early in the new century, Everett's own family and the family group into which he married each joined a "hiving off" of their colonial settlements into new ones in central New York. This was a passing frontier and a stepping stone toward the great explosion westward that took them to the southern fringe of Michigan soon after it became a state.

By that time Everett had evolved from a farm boy to an enterprising Connecticut Yankee, picking up an engineering education as he went. He learned the construction business in the blasting and excavation of the Erie and connecting canals, with their locks and resevoirs. He observed the invention of modern cement and became the supplier of it to the builders of New York City's new Croton waterworks.

But, after some four years on the Hudson, he was restless again -- his inherited westward urge prompted him to lead another "hiving off" to Jackson, Michigan. Jackson was chosen because the new railroad intensified growth and offered opportunities for major trading activities.

Another four years and Everett's dealings up and down the rail line exposed him to a new excitement -- America's first mining rush. It mobilized in nearby Detroit, but launched off into the fringes of civilization, where only Indians, missionaries and fur traders had lived before. By the time Everett joined the rush, trained scientists and financiers from the East had become involved.

Most of his relatives had settled down as farmers and Everett had to enlist a dozen neighbors for an expedition. He himself had to pick up new skills, but he was always a quick student. Otherwise he could not have survived on Lake Superior without the lore accumulated by the generations of French Canadian fur traders. From them he also learned how they dealt successfully and fairly with the Indians.

It was thus that he made a great discovery -- the greatest iron mine yet found in this country. Its tonnage was of such magnitude that it could not succeed before a major canal could be built and a railroad imported from "down state". The practical Yankee saw at once that the enterprise was beyond the means of his little group. First he urged the sharing of the finds, and the choice fell on some capable Cleveland explorers. Even so Everett and the expanded Jackson group could not finance the project and Everett early turned to a more feasible corollary -- to develop a port city to serve the lifeline to the mines. As he was still cautious, he and his "brother" (his brother-in-law Charles Johnson) shared this investment also with the Cleveland investors.

The story continues about how his several businesses flourished, taking different shapes over more than twenty years. During this time they survived the great fire that left them and all Marquette in ashes. The nationwide Panic of 1873, however, crushed his bank, but by this time he had raised his family and had had a fulfilling career. Although he had recovered with alacrity from the first of these stunning blows, old age and failing health prevented any comeback from the second.

After the death in 1882 of his wife, who had herself been a pioneer in these ventures for fifty years, Everett was more alone. However, he was comfortable in his daughter's family among those who loved and respected him. Thus he survived while his achievements were quite forgotten by the new generation that was living comfortably in the homes he had laid out and with the prosperity that the mines brought. Most important to Everett was the legacy to his descendants of honor and integrity in good times and in bad.

### ANOTHER WORD OF INTRODUCTION FROM THE AUTHOR

Many years ago I was struck by the confusion and contradictions in virtually all accounts of the discovery of iron and of the first iron mine in the Lake Superior country. The more accounts I gathered the worse the problem became.

A few were deliberate fabrications. Some arose from a competion (which I have tried to avoid) as to who was "first" in various achievements. Others were admitted fantasies -- not spoiling a good story for the sake of the truth, as Justice Voelker explained in the case of Laughing Whitefish. His book,[2] however, led me and many others to the best source -- the case of Charlotte Kawbawgam contained in the sworn testimony and the judicial findings in Kobogum et al.v.Jackson Iron Co.[3] and related cases (cited to the official reports hereinafter).

This definitive source laid to rest many falacies, such as the legend of the fallen pine tree, which I am now told was fabricated (probably just for publicity appeal) late in the 19th century. Certainly there is nothing to support it in contemporary records.

# THE BIOGRAPHY

## Everett's Family

Philo Marshall Everett was born on October 21, 1807[4] in Winchester, Connecticut. His was a distinguished Colonial ancestry, founders of the towns along the Connecticut River and some of the towns that "hived off" from them.[5] They were somewhat impoverished after his great grandfather Everett was bedridden for years before his death in 1765, but the widow then moved the family to Winchester in the hills of Connecticut, where she built one of the first houses, farmed and maintained a tavern. Everett's grandfather took care of the farm as long as his mother lived, but his older brothers left to obtain professions-- one became a minister, the other a physician.

## New York and the Erie Canal

In Everett's infancy his parents and grandparents moved[6] to Vernon, New York, a small town about 20 miles west of Utica and established farms. But even at sixteen, Everett resolved to escape the drudgery that had trapped his father and grandfather. A great new world opened up nearby where the Erie Canal was literally cutting a swath though the State of New York and the history of the new Republic. It offered not just a job, but an education in new skills and technologies. Everett was quick to seize on all these opportunities.

He reminisced in an undated memoir (handwritten in his old age and reproduced verbatim in its original spelling and wording in several excerpts hereafter)[6]:

My Father was a farmer, and with his farm and plenty of hard work he made the two ends meet. He never got in debt — but it was to slow business for me. I was anxious to see some of the World,  but

I had no money to travel with. I conseved the idea of getting on the Erie Canal and see what there was between Albany and Buffalo . I ... had seen but little of the world. Father was always kind and consented that I might try my fortune if I wished. He gave a little more than money enough to get to the Canal.

The Canal had opened late in 1825 after eight intense years of construction. It was a triumph for De Witt Clinton and the greatest engineering achievement that the country had known. The work of planning, construction and utilization created a whole new body of technology that inspired a wave of internal improvements in many states.

The first thing the Canal did was to open up a stream of commerce, thus developing farms and manufactures along its route. In one direction, it made New York City the leading American port and, in the other direction, it opened up the fullness of the old Northwest Territory to settlement and division into new states. Several decades passed before the newly invented railroads could offer competition, but finally they restricted the new canals to heavy bulk freight.

Meanwhile, the Erie Canal carried all the freight and passenger traffic through New York State. The many low bridges needed to carry across the Canal the roads and the farms whose lands had been dissected by "Clinton's Ditch" made hazards for passengers and crews alike. Throughout his life Everett would say that a shout of "Low Bridge" would cause some people among his contemporaries to duck.

Operating on the canal was not easy. It began with steering the cumbersome barges. Many more skills had to be created -- passing other barges without fouling lines, easing through the many locks and finally standing to steer without being knocked overboard.[7]

Many were the technical innovations in the building of the canal. Labor was short. so Yankee ingenuity had to substitute. Existing techniques, like blasting with black powder, were inadequate. New inventions, machines to pull stumps, horse drawn scoops and better concrete were all used. All these Everett studied and learned. Most important, he became impatient with the sort of seaman's life that prevailed and was accentuated on the canal. The bargemen had a short season on the water and then were laid off to carouse for long winter months "on the beach".

The farm boy tells in his own words how he got his first jobs and rose ever higher:

I took the first offer, which was bowsman at ten dollars a month. I was perfectly green on the Canal, but was anxious to learn and sought evry oportunity to stear and I gained much good will by stearing for the regular Stearsman, when in fact I was learning for my own benefit and I was soon placed in that position. The Capt. was a quick tempered man and often changed his hands. I went with him two years. The first, I saw so many pretty things that cost only twenty-five cents that when fall came I had but little more than to take me home .

The next year I turned my back on all those pretty things that I did not actually need and, when we ware in Albany and night came, it was hurry for the theater, but some one must stay and take care of the boat. I volunteered and was warmly thanked for my generossity, but in the morning I was a dollar ahead, they having spent that or more.

The third year I was in command of a boat, but I saw that the Canal was no place for me to only work six months and idle the ballance. Yet I saved a little, but I wished to accumulate faster and could do so by constant work.

## Moves in New York State

As our young Captain still sought something better and was fascinated by the burgeoning industries, an opportunity opened up close to his old home:

> I was offered a partnership in runing a Sawmill, which I accepted for the purposes of learning to run a Mill. When the year was out. I began a Mill in Boonville on Black River, but after a short time I found I could manufactor lumber three times as fast as it could be sold and that obliged me to give it up.

As Everett outgrew this too, he observed with some satisfaction:

> But I had aded considerable to my substance and owned a house and village lot and ten acres of land nearby. It was here where I found and married my Wife.

The marriage was at Utica, N.Y. on November 19, 1833. She was Mehitable Evans Johnson, born in Concord, New Hampshire on December 8, 1815. She had the same Colonial ancestry and had also experienced a move with her large close-knit family from New England. This was an ideal marriage. For fifty years Mehitable was a full partner in Everett's affairs. Herself a pioneer, she undertook the hardest travels, such as the two trips to Lake Superior in 1849 and 1850.

**PLATE 1**
**Canals of New York State**
*(Courtesy of The Erie Canal Museum)*

# Canals of New York State

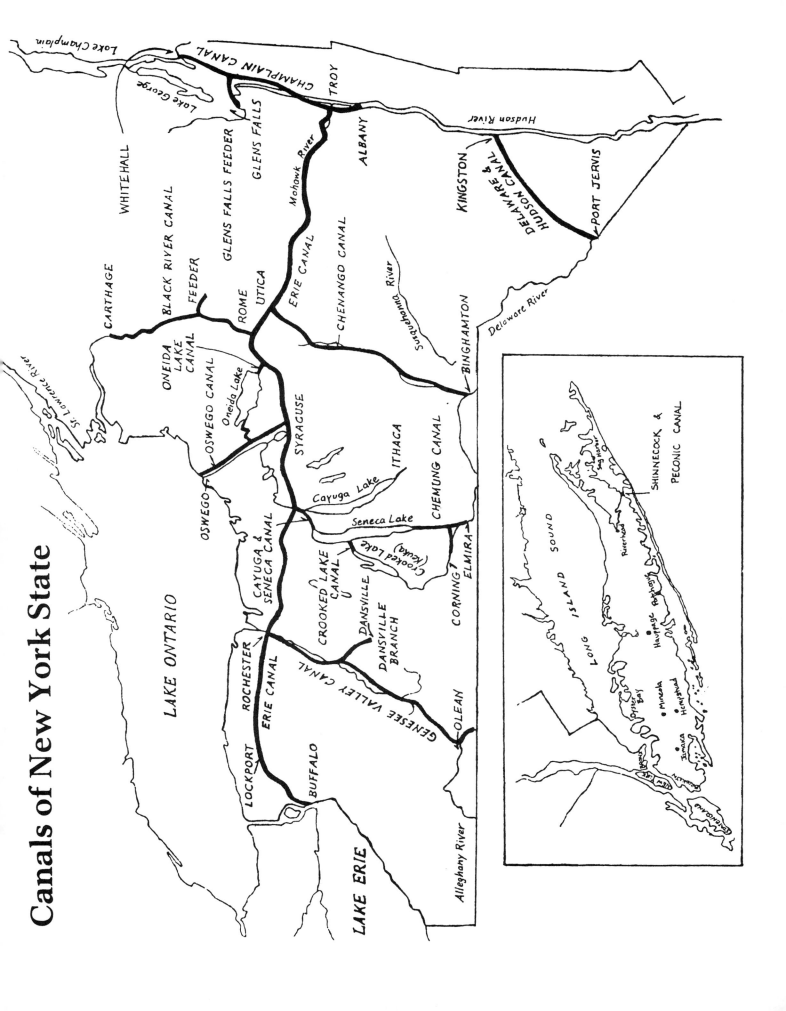

Everett became a full member of this family and soon its virtual leader, filling the vacancy left by the recent deaths of Mehitable's father, Moses Johnson, and her grandfather, Col. William Johnson, a Revolutionary War veteran, who had theretofore led the group in its several moves. Mehitable had three brothers, two sisters and several aunts and uncles. Most of them moved along with the Everetts, at least as far as Jackson, Michigan.

Everett tells how he moved into construction work, rising from being a "forman" into an operator: "I was then employed on the Chemung Canal as forman building looks, bridges, building[s] and a Resevoy [reservoir]." This was a branch canal, built to connect the Erie Canal with the Susquehanna River via Seneca Lake, Elmira, N.Y. and Athens, Pa. (See Plate 1.)

That was the best school I ever had. It learned me to manage men and teams, for we had a large number of both and I had to furnish all kinds of material for such work, both stone and timber.

After that Canal was finished (which took three years), I received an order to go down the North [Hudson] River, [at Whiteport] near Kingston and build up a Cement works capable of making from eight to ten hundred Bbls of Cement a day for the Croton Water Works — for Supplying in New York with water. After I had completed the works in building kilns and Mills, I took a Contract of Manufactoring, but it took two years to get the work in full blast. My Contract lasted two years untill the New York Croton water works were finished. My Contract was a prophitable one.

It was at Kingston that a second Everett child was born, Emma Eugenia, on August 15, 1839. (The first had died young.) Emma had, and passed on, a Delft tile of the Stoning of St. Stephen, that she said was taken from a "Holland Dutch" house there.

## Jackson, Michigan 1840-1845

Once the Croton contract was done, Everett held a family council. The frontier spirit that had led his family and the expansive Johnson family to New York was strong in all of them.

The lands in the Old Northwest had been filling up ever since the Erie Canal gave access to the Great Lakes. In the new state of Michigan there had already been a land rush and much of the best land in the lower tier of counties had been grabbed by speculators from the government at give-away prices. The real settlers had to buy from them at several times original cost, but the prospect was still attractive.

What particularly excited Everett was the way in which Michigan was aggressively pushing railroads from Detroit westward toward Chicago. The main line of the Michigan Central had already reached Ann Arbor and was reaching toward Jackson, a county seat that was doubling and redoubling a population that was only 500 in 1836.

Everett was too late to get a construction contract, as he hoped to do in view of his prior experience, but he could turn his hand to new opportunities. Accordingly, in just the way that Yankee ancestors had led new "hives" to swarm across western New England into New York, Everett, as the leader, led the Johnson clan to Jackson. Besides Mehitable and little Emma, there were her Uncle Frederick, Uncle David Porter and his wife, one sister (Helena Johnson Huntoon with her husband and three daughters, Matilda, Mary and Emmaline) and two brothers, Charles and Frederick, both with their spouses. Gilbert, a third brother, stayed on at the cement works on the Hudson for several years. Most of the members of this cohesive family continued to follow the Everetts and so will keep on appearing in the narrative.

Everett's memoir says "I then went to Jackson, Michigan in 1840 -- dealt a good deal in real estate, owned and run a Ware house while the Cars stopt there." "While the Cars stopt" at Jackson, there were indeed great opportunities to unload their freight and sell or distribute it. Now for the first time Everett was also a real estate operator, as well as a storekeeper and commission merchant. In the new community there were no established wholesaling or other distributors, and Everett developed still more talents as he imported necessities, like iron and glass, in Detroit and further east, warehoused them in Jackson and sold them all along the railroad.

The older relatives took to farming again, but the brothers, Charles and Frederick, and Uncle David Porter listed themselves as masons and had plenty to do building new homes.

Some biographers stated that Everett was in poor health in Jackson, but he said that he was strong and healthy until the Spring of 1868. He proved it by the strenuous life that he led up to that year. The only contrary evidence is his baptism on a sick bed on November 18, 1843 by St. Paul's Episcopal Church of Jackson, witnessed by Mr. Samuel Higby. Mrs. Everett was also baptized on a sick bed on February 25, 1844, witnessed by Dr. Backus.

Their three children, Emma (born August 15, 1839 in Kingston, N.Y.), Edward Philo and Charles Marshall (born February 26, 1842 and July 4, 1843 in Jackson) were baptized on April 28, 1844, with their parents as sponsors. The parents later sponsored their daughter Mehitabel Ellen (born January 11, 1845) in baptism on July 13, 1845, although the father must have acted by proxy. (At that time he was on the expedition described below.) They later sponsored their son, William Henry (born December 25, 1847, died October 22, 1849, all in Jackson) in baptism by Rev. D. T. Grinnell on April 22, 1848. Finally Catherine Eliza Johnson Everett (born in Marquette February 22, 1851) was baptized May 17, 1852, with her parents and Mr. and Mrs. Fred Johnson (her uncle and aunt) as sponsors.

The Everetts had not been baptized in the Congregational Church of their ancestors. Instead they became pillars of the Episcopal Church. The church in Marquette, of which they were founders, was named "St. Paul's", possibly for this church in Jackson.

## The Mineral Rush to Lake Superior

In 1845, while the copper rush of Lake Superior was at a high pitch, Everett became interested in it through his business contacts.

At this point, it is important to interject a short history of this country's first mineral rush. For at least four thousand years, Lake Superior was a prehistoric source of copper, and the first French explorers heard enough of it to excite their interest. Ancient man and the historic Indians used this native copper by merely hammering it into shapes with stone hammers.

After Michigan, including the Upper Peninsula (with much of Lake Superior), became a state, the brilliant Douglass Houghton was appointed State Geologist and went to the Upper Peninsula.[8] At the end of 1840 he issued an explosive report on the abundant copper and silver he found. A copper rush resulted, made possible by a treaty with the Chippewas ceding all the western part of the Upper Peninsula. (See Plate 2 for the ceded lands.) By 1843 several shiploads of prospectors had landed at Copper Harbor at the tip of Keweenaw Point. A government land office and fort were established there by 1844. A few copper mines were located with little real success, although sanguine reports by the illustrious Professor Charles T. Jackson[9], added fuel to the fires. In 1844 Jackson examined some of them and pronounced one of great potential. He followed with an article early in 1845, reviewing some that he selected "as undoubtedly valuable .....for profitable mining".[10]

This copper rush required appropriate Land Laws. Up to June of 1845, the War Department was granting permits to locate mineral deposits, first for nine square miles and then for one square mile. This practice was then stopped, but the permit holders were permitted to buy the land.

What was vital was a survey of the land. Starting in 1840, William Austin Burt, under subcontract with Douglass Houghton, began surveying townships six miles on a side, working westward from Sault Ste. Marie. By 1844, Burt reached the area of the present city of Negaunee, and on September 19th he went south from what is now called "Teal Lake" for six miles, establishing the eastern line of Township 47 North, Range 27 West. He routinely gathered mineral samples which he later turned over to Professor Jackson and is said to have remarked on the strong magnetic disturbances on the line. Among many other samples, he included iron ore from the first and second miles.

PLATE 2
Mineral Lands of Lake Superior

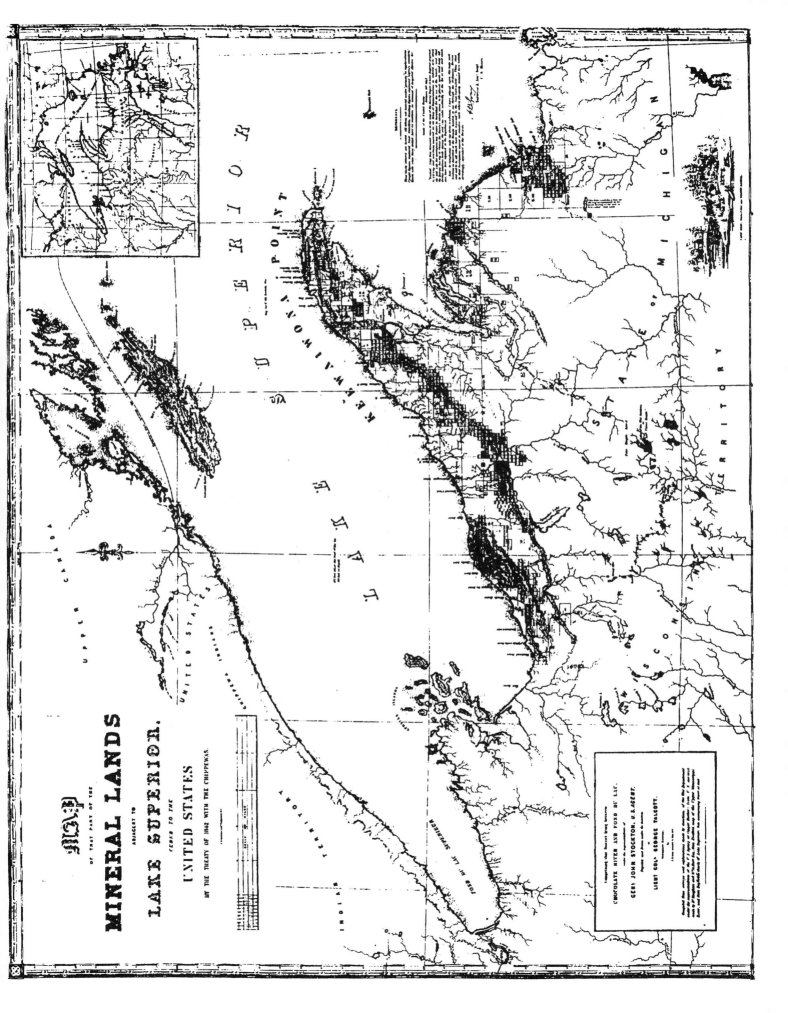

MAP
OF THAT PART OF THE

MINERAL LANDS
ADJACENT TO
LAKE SUPERIOR.
CEDED TO THE
UNITED STATES
BY THE TREATY OF 1842 WITH THE CHIPPEWAS.

Contemporary records of that year attest to Burt's lack of interest in the cause of the disturbances and much greater interest in his invention, the solar compass. When the ordinary compass went crazy, his compass proved itself essential to the surveying job.[11] Houghton made an examination of the area at the southern end of this line and made a note that the "metamorphic region will prove most important for its valuable iron ores".[12] There was no publication of this comment nor any report that winter that mentioned iron. In 1845, Houghton was lost in Lake Superior without having given iron any such stimulus as he had given to copper.

Professor Jackson also heard reports of iron ore in 1844 from an Indian chief near Sault Ste. Marie, Canada. In an article of January 20, 1869 in <u>The Mining Journal</u> of Marquette, Editor (later Governor of Alaska) Swineford described this as a report that a Mr. Barbeau had passed on and how Prof. Jackson, in the summer of 1845 sent Mr. Pray to investigate. They said that he found "a vast amount of iron ore" south of L'Anse, near the Wisconsin border.[13] A Mr. Stacy was also said to have found iron ore, although the localities "he visited are not definitely known". Mr. Swineford's article cast doubts on whether either of these was a discovery of any value.[14]

<u>The Expedition to Lake Superior — Summer 1845</u>.

Returning to Philo Everett some 500 miles away, we note that he knew nothing about iron or even the existence of these surveys. It was the copper rush that had already been going on for several years that motivated him.

The following is Everett's own account:[15]

In 1844 the copper interest of Lake Superior got to fever heat, especially in Boston, by the reports of Professor Jackson, of Boston.[16] A friend of his in

Detroit, whom I was then doing business with, gave me the history of Jackson's work in exploring on the shores of Lake Superior, and in the spring of 1845 I determined to visit Lake Superior and see for myself, if possible, what all that talk amounted to. I proposed to some of my friends to join me in a speculation of that nature, and I soon collected thirteen members.

The Johnson relatives, with the exception of Charles, were too well settled in Jackson to be at all interested in another move, and it was with considerable effort that Everett could recruit this group in the Jackson vicinity, supply the expedition and do everything else required in the short time available

They formed a company to be known as "The Jackson Mining Company," the original officers of which were: Col. Abram W. Berry, president; Frederick W. Kirtland, secretary; Philo M. Everett, treasurer, and to be in charge of exploration. The others were John Westron, William A. Ernst, Francis W. Carr, Solomon T. Carr, William Monroe, Fairchild Farrand, Edward S. Rockwell, John Watkins, S.A. Hastings and James Ganson.[17]

Everett wrote[18]:

I then sent to our Senator at Washington, Mr. Norvel, asking him to procure a number of permits from the Secretary of War, giving permission to locate a mile square each anywhere on the south shore of Lake Superior, for mining purposes and I believe that was the last day permits were ever issued, the transaction of the Secretary of War being declared illegal; but Congress legalized the act and gave permission for any person to locate a mile square on the south shore of Lake Superior by leaving a person in charge of the location.

I received our permits on the 19th of June, and on the 20th of June, 1845, I left my home in Jackson for Lake Superior. [Accompanying him were S. T. Carr, Rockwell and William Monroe.[19]] I bought my supplies in Detroit and took a steamer for Mackinaw, as there was no boat then running direct to the Sault as now [written in 1887]. I purchased a coasting boat at Mackinaw (See Plate 3.) and put it on board the "General Scott", a small side wheel steamer, making three trips a week from Mackinaw to the Sault. It was said no boat could go up the Sault river then, drawing over nine feet of water.

For some two centuries Sault Ste. Marie had been the key to travel beyond. The Sault was not a waterfall, but a set of rapids that required a portage, at least on the upward voyage. The fur trade on Lake Superior and nearly all of the native population had disappeared, but the copper rush had revived the Sault into a boom town

Everett continues again[20]:

I was somewhat surprised on arriving at the Sault to find such an immense warehouse for traffic with the Indians of the northwest. My first duty was to transport my coasting boat over the portage of three quarters of a mile and to ship the most of my supplies to Copper Harbor, that being copper headquarters.[21] We struck our tent at the head of the portage.

PLATE 3
Model of a Mackinaw Boat

Above the Sault, on Lake Superior, transportation was almost non-existent.[22] During the 1844 season, the War Department moved two companies of soldiers to the end of Keweenaw Point and built (with materials shipped from Lake Huron) a complete fort, Fort Wilkins. In doing this, they tried to charter the schooner Algonquin and the "so-called brig" John Jacob Astor of 112 tons.(See Plate 4, showing a brigantine entering Grand Island Harbor.) Unfortunately, the latter was wrecked at Copper Harbor in a September gale.

This was only one of the perils that put the miners in desperate danger of starvation during the winter. Experienced copper miners, like John Hays, were fighting to compete with the Army and charter the Algonquin. The Army was pushing to have two small schooners and finally the Independence, a "propeller" and the first steamboat on the lake, hauled around the Sault in 1845, but none of this was done in time to help Everett on this first trip.

The Jackson party had the good fortune to be advised and helped by an old hand, Samuel Ashman.[23] He seemed to favor them with advice over all the other copper seekers. The Jackson Mining Company had raised $650,[24] which seems to have been an adequate provision for the expedition. While they were not among the penniless prospectors, their total capital would not have brought three mules at Copper Country prices, and they were not in the same league as the Hays group, which started with $25,000 of Pittsburgh and Boston money.

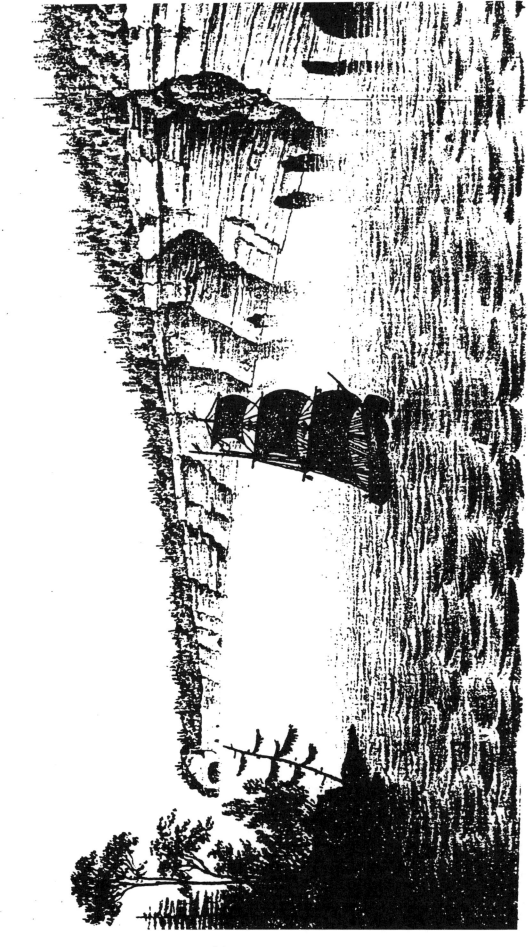

PLATE 4

Brigantine *John Jacob Astor* at Entrance to Grand Island Harbor

24

Ray A. Brotherton, writing in 1945,[25] now puts an interesting story in Everett's mouth that seems consistent with what follows:

> While dealing with [Chief] Ma-dosh, ... I became acquainted with Tipo-Keso (the Full Moon), who was the daughter of Man-gon-see (Small Loon), a brother of Chief Marji-Gesick, who claimed the land in the vicinity of the Carp river and Teal lake as his hunting ground. [See Plate 5 for Marji Gesick and other Indian chiefs.] She had just come from there and told us of a mountain of rock or mineral too hard and heavy for the Indians to use. She gave us careful directions of the location and the distance of the trail from the mouth of the Carp river, where it enters Lake Superior, to her uncle, Chief Marji-Gesick's camp on Teal lake, and assured us that he would take us to this mountain of Heavy Rock. [Compare this account with Professor Jackson's account[26] of being given in the summer of 1844 "a fine specimen of specular iron ore".]

> Mr. Ashman told us that the navigation on Lake Superior in small open boats was very dangerous, and, in case of shipwreck, the water even in the summer was so cold as to be numbing, causing cramps and few persons could swim very far. Storms rise very quickly and with very high winds, and he advised us to add block and tackle with rope of ample length, and strength to enable us to draw our boat out of the water each evening in case of sudden storms.

## PLATE 5
### Marji Gesick and other Indian Chiefs

*(UPPER ROW, left to right:)*

1. Rev. J. H. Pitazel, missionary located at L'Anse, who in 1848, married Anna Williams of Grand Island to Aaron Powell.
2. Mon-go-sid, *(Loon Foot)*, Marquette Indian.
3. Mon-gon-see, *(Little Loon)*, Marquette Indian and brother of Chief Marji-Gesick.
4. Marji Gesick, *(Moving Day)*, Chippewa Indian Chief of the Marquette Indians whose tribe numbered about thirty. He made his home on Teal Lake near the outlet, now in the city of Negaunee and is the Indian who showed Philo Everett the Jackson Iron Mountain. He died in 1862 and is buried in an unknown grave in Marquette County.
5. Charley Kaw Bawgam, last Chippewa Indian Chief, son-in-law of Marji-Gesick and who became Chief upon the death of Ma-dosh. He made his home at Marquette. He died in 1903 at the age of 103 years and rests in a grave on Presque Isle.
6. As-sin-nins, *(Little Stones and Pebbles)*, L'Anse Indian for whom Father Baraga named his mission at Keweenaw Bay.
7. Lo-kwa-da, *(The one who creeps, who does not walk)*, Marquette Indian.

*(LOWER ROW, left to right:)*

8. Ma-dosh, the first Chief of the Chippewa to whom the U.S. Government bestowed a medal. He was an expert boat and canoe builder, living on Chief Island at Sault Ste. Marie.
9. Kish-kit-a-wage, *(The Indian with an ear cut off)*. He made his home on the shore of the Bay where Munising is now located.
10. Matchi-kwi-wis-ens, *(Bad Boy)*, L'Anse Indian.

Philo Everett's Recollections in 1887[27]gave his account
of his arrival at the Sault:

> The next thing in order was to procure a
> coaster, one that was familiar with the lake. The
> thought of coasting along the rocky and desolate shores
> of Lake Superior, not knowing at any time what we were
> to meet with next, was not a pleasing one, especially
> in passing the Pictured Rocks. It was well known that
> there were long stretches of coast there where no boat
> could land and that Lake Superior often got very angry
> in a few minutes. (See Plate 6 for a Mackinaw boat off
> the Pictured Rocks.)
>
> Louis Nolan[28] was recommended to me as the
> best man for that purpose in the Sault. He was a
> large, stout man, well acquainted with the lake and all
> the northwest....He was a little over six feet high,
> well proportioned, a Frenchman with a slight mixture of
> Indian blood, with an intelligent countenance and
> pleasant address, and very polite. I made my business
> known and inquired if he was well acquainted with the
> lake. He replied that he had coasted the length of the
> lake many times, on both sides, and also had traveled
> many times to Hudson's Bay and had been employed by the
> fur company for many years as a clerk.
>
> Now we think of a clerk as one sitting in an
> easy chair in a warm office, writing at a desk; but a
> fur company's clerk is quite a different thing. He
> must be able to write a fair hand, be a good
> accountant, and be able to take a ninety pound pack on
> his back and travel all day from one Indian camp to
> another, collecting furs and living entirely on wild
> meat, mostly rabbits, for the fur companies only

PLATE 6

A Mackinaw Boat at the Grand Portal of the Pictured Rocks

supplied bread food enough to last their clerks to headquarters. The ninety pound pack consists of Indian goods, and that is the standard weight of all fur companies' packs. Knowing these facts before, I was at once satisfied he was the man for us. He made a proposition for the season to pilot, pack and cook for us. The bargain was then concluded, but he wanted two days in which to prepare for the summer trip. It was granted.

He now remarked: 'You say you are going to Copper Harbor for copper ore; you don't want to go to Copper Harbor for ore, there is plenty at Carp River. There is more ore back of Carp River (now Marquette) up at Teal Lake than you can ever get away -- two mountains of it -- only two or three miles apart.' I inquired what kind of ore it was. 'Don't know much about ore' and, having a few specimens of ore with me, I spread them out and requested him to point out the ore like the ore at Teal Lake. He shook his head, putting his finger on a piece of Galena lead ore saying that was the most like it, but that wasn't it. 'It looks like rock, but it wasn't rock, several bowlders lay beside the trail, worn smooth, and shined brightly.' 'When did you see this ore last?' I inquired. 'Thirty six years ago I went from Carp River to Menominee with some Indians, and never having seen anything like it, I distinctly remember it'. 'How old are you?' 'Most sixty.'

PLATE 7

Indians fishing in the Rapids at Sault Ste. Marie

30

His description of the ore two or three
miles further on was equally surprising. The trail ran
along the north side of a bluff, fifty feet high, of
solid ore. This description greatly surprised me, for
I learned he was a Christian man of the Roman Catholic
faith, was perfectly truthful and reliable, never used
profane language and never got drunk. What could it
be? It was not copper, that was evident, for I showed
him copper specimens, and that was not it, as he termed
it. I had never heard of iron in this district, and
therefore thought nothing of its being iron.

In his continuing Recollections[29], Everett continues
with several more vignettes of the Sault:

Now, as I had two days to wait, I took a stroll about
the town. Passing down the portage, I noticed several
canoes in the rapids, two Indians in each canoe,
standing erect as steadily as if on land. I watched
them for several hours, for I had never seen or heard
of such a way of taking in fish. (See Plate 7.) The
man in front soon dipped in his scoop net and took out
a large whitefish. It was strange to see how that
frail bark canoe could be shot into the foaming rapids,
as white as milk, and could be managed by that Indian.
The forward one, when he saw a fish, would lay down his
setting pole, take up his scoop net, dip up his fish,
and again take up his setting pole with surprising
ease, and the canoe would again be shot into the
foaming rapids still further. Few white men could
stand erect in this canoe a single moment. I sat on
the shore for a long time, scarcely thinking of the
passing hours.

I next visited the fort, a beautiful site for a city --- such a handsome plat of ground on every side.

Not far from the fort was the Baptist mission, under the charge of Rev. Mr. Bingham, a very pleasant gentleman. He told me he went on board of a schooner at Buffalo, with his family, bound for the Sault, if I remember rightly, in 1833, to take charge of the mission. He had a lovely family -- his girls were like roses in a wilderness. He told me much of his labors with the Indians; he thought he had done them much good, and I had no doubt of it. He had taught them to read and write.

But the two days were now wearing away. Louis reported at our tent for duty with his pack of blankets and tent cloth, together with a shot gun, having the appearance of being manufactured in Queen Anne's time, but it was a deadly weapon, dangerous at both ends, as one of our party could testify a few days afterwards. He ventured to fire it at some game, was knocked sprawling on the ground, and went with a lame shoulder for many days. He said it kicked like a mule. No one of our company had the courage to fire it afterwards during the whole summer.

Everett's Recollections[30] continue:

We were not long in finding out that we had made a wise choice in our coaster. He knew every point and every stream that entered the lake. When it came time to camp, he would run the boat ashore at the mouth of a stream where we could catch all the speckled trout we wanted for supper and breakfast. He never left his seat on the stern of the boat. When we sailed, he steered, and when we rowed he paddled and steered.

He was supplied with trolling line, as well as gun. The first day I said to him, 'Can't you catch some trout by trolling?' 'No trout here; too much sand beach,', he said; but one day as we were passing a rocky point he took out his trolling line, saying, 'May be we can catch trout here.' He threw out his line and a big trout took it before the hook was twenty feet from the stern of the boat, and I saw several others after it. He took in several fine ones in a few minutes and went to winding up his line. I asked him to let me take it and catch a lot. I shall never forget the look he gave me, saying; 'What you want of them? You have now more than we can eat; do you wish to waste the Indians' food?' That was a break-down argument. I admitted he was right, and said no more.

Brotherton's account includes some more reconstructions of the voyage:

We were soon on our way with our sails filled with a fair wind which enabled us to reach Whitefish Point in ample time to prepare for a night's sleep before dark. It was here Indians came to us with large fresh-caught whitefish which we bought for a small sum and had them for our supper.

In the night we were wakened by our tents tumbling down upon us. We crawled out in to a furious wind and the thundering roar of the dashing waves, that had nearly reached our supplies, and our boat was at the water's edge instead of being 50 feet away as we had left it the night before. Twice during the night the boat had to be hauled further back, and our tents were twice re-pitched.

PLATE 8
Chapel Rock, about 1851

34

When daylight came the wind was blowing hard
and the waves too high for us to attempt to launch our
boat, so we remained until the second morning, when
upon arising we found the lake smooth, the morning
bright and sunny. Our boat and the supplies were 200
feet away from the water, but we were early in the boat
and away. No wind came with the sun and we had to take
our turn at the cedar oars, and, big and heavily loaded
as our boat was, we were able to reach Grand Marais
Harbor before night.

Here again we were wind bound until the
second day, when even though the wind was rather fresh
we decided to chance it, passing the famed sand dunes,
being immense banks of high and abrupt sand, rising to
a height in some places to 200 feet. Before us now
appeared one of nature's wonders, the art gallery of
Lake Superior, the Pictured Rocks, but our Admiral was
not looking at the towering sandstone cliffs, but was
casting fearful eyes at the large waves that threatened
to engulf our small boat at any minute as the wind had
increased to a gale. In the distance we could see the
Grand Portal and hear the thundering of the waves as
they entered the enormous caves underneath it, and as
Chapel beach hove in sight, ... [we] decided to beach
our boat, where, after much difficulty and hard work,
we pulled back some 100 feet from the surf and wet and
weary made our camp.

.....The Grand Portal was one of the main features of the Pictured Rocks that extended for 18 miles along the south shore of Lake Superior, from Grand Marais on the east to Munising on the west; in places rising in sheer precipices to a height of 150 to 200 feet above the water line, showing a succession of sandstone sculptures whose effects are heightened by the brilliancy of the coloring -- alternating yellow, blue, green, brown and grey in a bewildering series of shades, with many of the rocks given such names as 'Grand Portal,' 'Doric or Chapel Rock,' 'Profile Rock' 'Indian Head Point,' 'Battleship Point,' 'Sail Rock,' 'Lover's Leap,' 'The Cascade' and 'Miner's Castle'.(See Plate 8.)

The storm finally abated allowing us to continue our journey and we rounded Miner's Castle and entered the east channel with Grand Island looming up to our right. We soon reached the south end of the island with its well protected harbor, where on landing we were met and welcomed by Abraham Williams, his wife, and 13 children.

Williams had a number of log cabins, quite comfortable, with bunks and a large stone fireplace in each, one being a trading store in which were such goods as Indians wanted. (See Plate 9.) The store was kept locked, being opened only when the Indians came with furs to trade, or without them to get necessities, credit being given against next year's catch. There were no white people living nearer than Sault Ste. Marie, more then 100 miles away, and the Williams were glad to see us and do all in their power to make us comfortable.

PLATE 9
Williams' Cabin
*(Moved to Munising)*

## Finding the Mysterious Mountain

Finally the party reached the mouth of the Carp River, where they had been told to meet Chief Marji Gesick. Now we come to the heart of the story about the discovery of the "iron mountain" and find several inconsistent versions. We must, I believe, lean toward the most contemporary accounts. Furthermore, the matter was the subject of considerable litigation that went three times to the Supreme Court of Michigan in 1882, 1883 and 1889[31]. While the exact circumstances of the discovery were not crucial to their decisions, they did have the benefit of some 30 pages of sworn testimony by Everett, including some hostile cross-examination. The summary that the court gave of the discovery[32] was as follows:

The Jackson Company sent up persons to explore and secure mining lands. These gentlemen heard there was a valuable iron deposit, of which one Lewis Nolin (sic), a half-breed, knew something. Not being able to find it, they were recommended to go to L'Anse, and find an Indian chief named in the record Marji Gesick, in whose territory the land lay, and who could take them directly to it. Without making any definite bargain about his pay, they told him he would be rewarded; and he went with them, and took them to a remarkable and valuable iron bluff or mountain, since known as the 'Jackson Mine', near Teal lake.

In another decision[33], they said much the same thing:

Some members of the company went up to Lake Superior and prepared to explore. Marji Gesick, being with his people at L'Anse, which is considerably further up the lake, and being regarded as well qualified to give useful information concerning mineral deposits, Mr. Everett, one of the company's agents, sent for him to assist them. Marji

Gesick came over and showed them the iron mine in question; and the location was made by placing the center near the principal outcrop, and surveying a square mile including it.

Note that the court's summary conforms with the account of the meeting at the Sault of Philo Everett and Tipo-Keso, the daughter of Man-gon-see and the niece of Marji Gesick. He did exactly what she recommended, although her uncle was not at Teal Lake but at L'Anse. His Recollections do not emphasize that Nolan could not lead him to the "ore" that he had described at the Sault, but Everett testified[34] that he landed at Carp River (where Marquette is now located) but, when then asked if he became acquainted at that time with Marji Gesick, he said, "I did, in the course of the summer; not then; he was not here; and this was a month or so after." His testimony at the trial in Marquette goes on as follows:

I employed him soon after that to show us the Jackson mine.

Q    to show you the iron deposit now known as the mine of the Jackson Iron Company. A    Yes, sir.

Q    Did he show to the company that mine: A    Yes, sir.

Q    At the time that he showed you that mine, was there any apparent ledge of iron there in sight on the surface? A    Oh, yes, sir.

Q    Well, how much of a ledge was in sight? A Well, it showed itself all the way from the foot of the mountain, as we called it, in different places, as we passed up clear to the top; but there was a ledge at the bottom some ten feet, I should think, high, and perhaps more.

On cross-examination, Everett was shown a letter (quoted below) to his brother-in-law, Gilbert Johnson.    He

40

pointed out one mistake in it, in that he left Jackson on the 23rd of June (not July) and "was up here on the 4th of July." The cross-examination then continued in this way:

> Q    You say that you met Marji Gesick at that time? A    I met him that summer.
>
> Q    Where did you meet him? A    I met him first at L'Anse.
>
> Q    Did he come to this country with you or for you? A    Yes, sir, he came here for me.
>
> Q    He came here at your request? A    Yes, sir.
>
> Q    Did you go into the woods with him? A    No, I sent two men with him.
>
> Q    Who went with him? A Carr and Rockwell

Neither did Marji Gesick go the full distance. Everett and Nolan were present, however, when the site was formally claimed on September 20. An 1873 report describes it as follows:

> The actual discovery of the Jackson location was made by S.T. Carr and E.S. Rockwell, members of Everett's party, who were guided to the locality by an Indian chief, named Manjekijik. The superstition of the savage not allowing him to approach the spot, Mr. Carr continued the search alone, resulting in the discovery of the outcrop, which he describes as indicated in Mr. Everett's letter. Previous to the discovery he was led to suppose from the Indians' description, that he would find silver, lead, copper or some other metal more precious than iron, as it was represented and found to be 'bright and shiny'.

The site was about 3200 feet west of the section where Burt's party had discovered ore the year before. As noted below, this was registered as Mining Permit Number 593, one

square mile in size, issued to James Ganson and purchased at
$2.50 per acre from the U.S. Government.[35]

The marking of a pine tree and the measurement to Teal
Lake, the supposed landmark, all became quite important later.
Further detailed testimony was given as follows:

Q. In what way was this mine designated -- marked
out so as to locate it? A. Well, there was a certain pine
tree there by the ledge that we marked, and then ran to the
south-east corner of Teal Lake, and it was bounded by
marking that pine tree the center.

Q. The center of the section? A. Yes, sir.

Q. What was the size of the pine tree about at
that time? A. I could not tell; two feet, I suppose,
thereabouts.

Q. Was it marked on all four sides? A. Yes,
sir.

Q. Do you remember how it was marked? A. No, I
could not tell; it is some time ago, and that is something
that I never charged my mind with.

Q. That was to be the center of the mile square?
A. Yes, sir.

The following testimony proves that the Jackson party
had never heard of the Burt survey:

Q. At the time you located this permit was the
land surveyed by the general government? A. Well, it was
townshipped, [i.e. laid out in squares six miles on a side]
but we didn't know it then.

42

Q. How long after was it laid off into sections [of one square mile]? A. It was in 1848 or 9; it seems to me that it was '49 that it was sectionized.

Q. After it was laid off into sections what was the description, if you can recollect, of the section that covered this land that the permit covered? A. Section one, town forty-seven, twenty-seven.

Q. Did the sectioning come near the land marks that you had marked out? A. Well, all I know about that is what the surveyors said.

Q. Have you ever been on the section after it was laid out in section line? A. Oh, yes.

Q. Were you at the center tree after that? A. Yes, sir.

Q. Did you ever notice whether these section lines, as run, came to about the boundary of the permit, as located? A. No, I never followed the lines at all.

[This testimony is very relevant to the events that occurred at the Mineral Agency office at Copper Harbor.]

Q. Where did this big pine tree, that you have mentioned, stand with reference to the mine as now located? A. I don't know as I can describe it to you so it will be understood; it was right where we first commenced to mine.

Q. Up on top of a little hill, wasn't it? A. No, it was not on top of a hill; it was some ways up beyond where their office is now.

Q. Well, which way — to the west? A. Yes, sir, to the westerly.

43

Q. South-westerly?  A.  I could not say; it was westerly.

Q. You know right were it stood, don't you?  A. Oh, yes, sir; the Indian trail went right over the mine and right past the tree.

Q. Do you know what was done with that tree?  A. I heard Merry say that he had cut it down and sawed it up and made a desk of part of it, and he showed me the desk in the office; that is all I know about it.

None of this gives a complete description of the Iron Mountain. It must have been very impressive because the Indians held it in such superstitious dread. The iron ores and the jasper sparkled brightly when the sun struck them. Everett's 1887 Recollections[36] give this description:

On arriving at Teal Lake, we found the ore.... There lay the boulders of the trail, made smooth by the atmosphere, bright and shining, but dark colored, and a perpendicular bluff fifty feet in height, of pure solid ore, looking like rock, but not rock, and on climbing a steep elevation of about seventy feet, the ore cropping out in different places all the way, we came, at the top, to a precipice many feet deep. Hundreds of tons of ore that had been thrown down by the frost lay at the bottom. It was solid ore, but much leaner than that on the other side. From all that could be seen, it seemed that the whole elevation for half a mile or more was one solid mass of iron ore. No rock could be seen, and all that visited it came to the same conclusion, until the mine was fairly opened. By measurement, the outcrop was found to be three quarters of a mile southwest of the southeast corner of Teal Lake. [A historical marker now stands at this corner.]

[An] other outcrop, two miles further on, was a beautiful sight. On the north side of the hill it was a perpendicular bluff of about fifty feet of pure iron ore and jasper in alternate streaks, but more jasper than iron.

Another small outcrop appeared a mile further on, for several years known as the little location (now known as Lake Superior mine). With all its beauty, that high bluff proved worthless; but the Cleveland mine, near by, only a few rods from its base, was soon discovered, and proved one of the best mines in the country.

In the testimony and then in the narrative there is just a little more characterization of what the mine looked like.

Q. Did this Jackson mine prove to be a very large deposit of iron ore? A. Yes, sir.

Q. You say that the ore cropped out of the side of the hill at the time it was first discovered. What was necessary in the way of mining to take out the ore at first? A. Oh, simply quarry it out.

Q. Simply blast it off and break it up? A. Yes, sir, the same as you would go into any stone quarry.

As Brotherton described it:

Here were to be seen thousands of tons of iron ore exposed above the ground. We made arrangements with the chief for a number of his men to pack about a ton of the iron ore over the trail to our main camp at the Carp river where he could load it into our boat.

45

It would be seen that all this conflicts with the much later popular legend to the effect that the Burt party found iron under a fallen pine tree and that Chief Marji Gesick learned about this and took Philo Everett there. The Marquette Centennial booklet of 1936 poked fun at this theory.[37] & [38]

There is also a picture of a mine captain sitting on a pine stump that is supposed to be this fallen pine tree. However, Mr. Mathews[39] told me that this was the marked tree above the Jackson mine, cut down, as described in the Testimony. He says that "Billy Pick," the mine boss, used to sit on this stump. He was called "Billy Pick" because during the lunch break he would select a miner and tell him that he had to pick faster during the afternoon or he would not be there tomorrow.

It has also been pointed out that the contemporary account of the "mountain of solid iron ore, 150 feet high... as bright as a bar of iron just broken" does not sound like a deposit under a fallen pine tree.


Copper Harbor

As all miners know, the most important thing was now to register the claim. The Government land office was on Porter's Island at Copper Harbor over a hundred miles away. Apparently they went there immediately, although they may have attempted another location for copper. Locating this iron mountain was the primary problem and Philo Everett described it, first in his Recollections[40] and then in his testimony, as follows:

On arriving at Copper Harbor, I found the government mineral office on the island opposite the harbor, which in fact formed the harbor. (See Plate 10) The white tents on the island appeared like an army encampment. Presenting my permit[41] and description, the officer looked it over, saying, 'Where is Teal Lake? It is not on my map.' I told him the Indians called it twenty five miles southwest of Carp river, and it took us a day and a half to go there. That was all I knew about it. He said to me: 'Mark the lake on the map.' I refused to do so, saying that it might work us an injury, as it was pretty certain to be wrong. He measured off twenty five miles on his map and marked out Teal lake with our permit on the south side, as given in the description. (See Plate 11.)

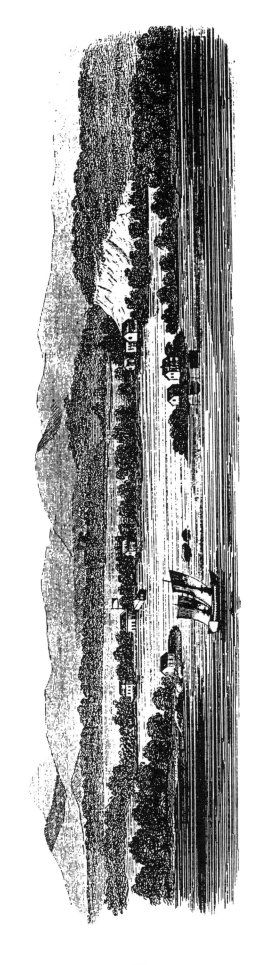

PLATE 10
Bird's Eye View of Copper Harbor, ca. 1845

48

This is the way he testified:

Q. Did you have seven permits with you at that time, or more or less? A. Yes, sir; I think I had more.

Q. You say that one permit was located upon that property? A. Yes, sir.

Q. Who located it? A. I located it, or my company, or some of them.

Q. What do you mean by locating? A. When I say locate, I mean go to the office and locate it on the map.

Q. That is what you mean by locating? A. Certainly.

Q. Where was the office? A. At Copper Harbor.

Q. What office was there? A. The mineral office.

Q. Mineral office of the War Department? A. Yes; the War Department kept a mineral office at Lake Superior.

Q. Did you go there and locate this permit upon that property? A. Yes, sir.

PLATE 11
Detail of map attached to
*The Mineral Region of Lake Superior*
(1846)

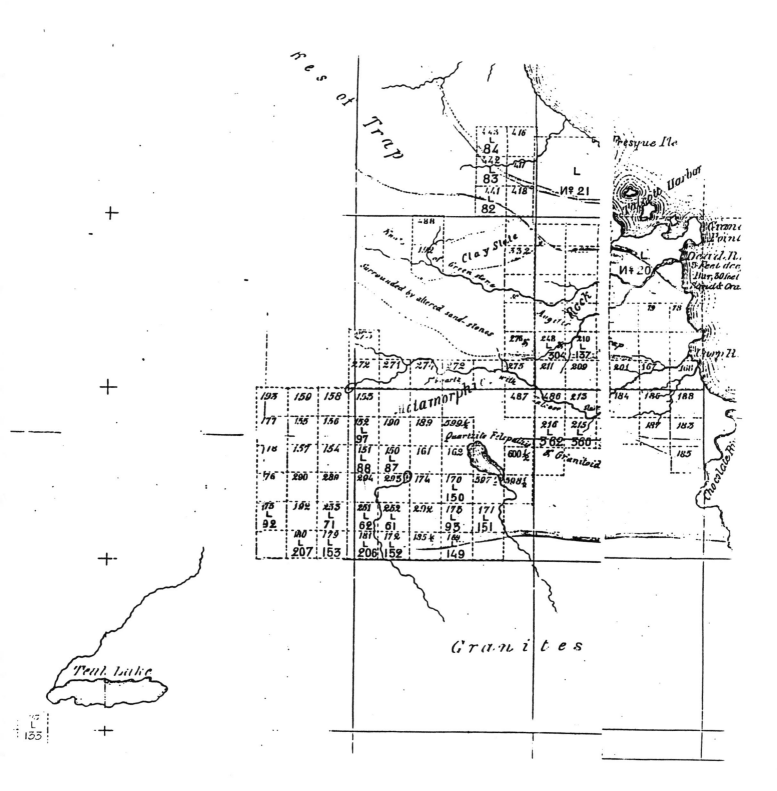

Q.  What was the process of locating it?  A. Well, we handed in our permit to the officer, and it was marked on the map.  It was not always marked in ink; it was marked with pencil.  They gave us that privilege, to mark it with a pencil, and then gave us the privilege to go and examine the land, and if we didn't like it we could lift our permit, as we call it, and they would rub out the marks.

Q.  Did you that summer locate other permits?  A. I think our men did after I left;  I left some men up here.

Q.  Well, did you?  A.  No, I think I did not myself.

Q.  Do you remember what town and range this section was supposed to be in -- this mile square that you located with this permit?  A.  I know what town it is in.

Q.  Do you know whether you knew then?  A.  No, I did not; I did not know that there was any survey there then.

Q.  Didn't the map show?  A.  No, sir.

Q.  What did the map show -- just an unbroken waste of country?  A.  Well, pretty much -- that ain't correct exactly; when I went there to the office the section was bounded by Teal Lake; the officer said, 'Where is Teal Lake?  Said I, 'I don't know; the mouth of Carp river is supposed to be twenty-five miles south [sic] of that; that is what the Indians call it,' and he located it; he wanted me to put it on the map; I refused to do it, and he put it on the map that distance, and it was 12 or 15 miles east [sic] of where Teal lake is, but he bounded the location by Teal lake.

Q.  You say that you located the permit at Copper Harbor at the mineral office?  A.  Yes, sir.

Q. At the time of locating it did you hand in to the mineral agent a description of the land that you located -- a written description of the land? A. I think we gave a written description of the location.

Q. Let me read to you this description and ask you if you recollect it as the description: 'Commencing at a spruce tree about eight inches in diameter standing on the bank of the south-east corner of Teal lake, or Shin-bi-hs-a-ga-mut lake, as called by Louis Nolin, a half-breed guide and former sheriff of Sault Ste. Marie, then and there present about the 20th day of September, A.D. 1845, from thence a south-west course one mile to a bearing pine tree about two feet in diameter, marked B.T., which letters are cut in on four sides, and which tree is intended to be the centre of the mile square tract desired to be secured in this location in the name of James Ganson, of Jackson county, Michigan, on which pine tree is also marked on the proper sides E. W. N. S. intending and indicating the corner on which to measure a half mile on each side from said tree, and at the end of which four lines is a blazed tree' You say that was the description of the land that you handed in? A. Yes, sir. I should think so.[Tr. p.121 shows "Shinbilisagamut Lake".]

In 1846 Jacob Houghton,Jr. published a report on The Mineral Region of Lake Superior, including a "List of Locations and Leases up to July 17,1846 .....Accompanied by the Corrected Map of the Mineral Agency Office...." [42]. The list of locations includes "No. 593, James Ganson, Jackson, Mich. and Lease 133, dated Dec. 12, 1845".(See below). A detail of the map (Plate 11, described above) shows a small oval lake at the corner of Townships 47 and 48 and Ranges 26 and 27, the source of the Carp River. This is the true location of Teal Lake, although it is too small on the map. About six miles to the west and twelve miles to the south (in an unsurveyed area) appears "Teal Lake", about four miles wide by one mile north and south and, south of its western end is a one-mile square, marked "593 L 133". This is apparently the way the officer "marked out Teal lake with our permit on the south side", as Everett testified.

The testimony continues:

Q. Did you ever see a certificate signed by the superintendent of mineral lands, John Stockton, certifying that the location which has been described was made by James Ganson, and did not interfere with any other location made? A. I saw one here the other day, I believe.

Q. Did you see one about the time the location was made? A. I could not remember that.

Q. Let me read this and see if you have any recollection of this — this is attached to the description: 'I have carefully examined the description of the tract above applied for and find that it does not interfere with any other location made under proper authority, and I recommend that a lease be granted as prayed for. Dated Washington city, this 5th Day of December, 1845. John Stockton, superintendent of Mineral lands.' Did you ever see that? A. I could not say that I did, because Kirtland was secretary and it would go to him, and I would not be likely to see it unless he showed it to me. I might never have seen it.

In 1847 the United States Senate caused to be published a large map, The Mineral Lands Adjacent to Lake Superior, (Plate 2) together with an "accompanying report".[43] It was based in part on "U.S. surveys made by Dr. Houghton and Wm. A. Burt, Esq." (as shown on Plate 11) and hence repeated the location of Teal Lake shown above. (This map is herein called the "Gray Map".) However, there was added beside the lake the following legend (See Plate 12):

**PLATE 12**
Detail of Plate 2,
showing Location No. 593
and Teal Lake mislocated.

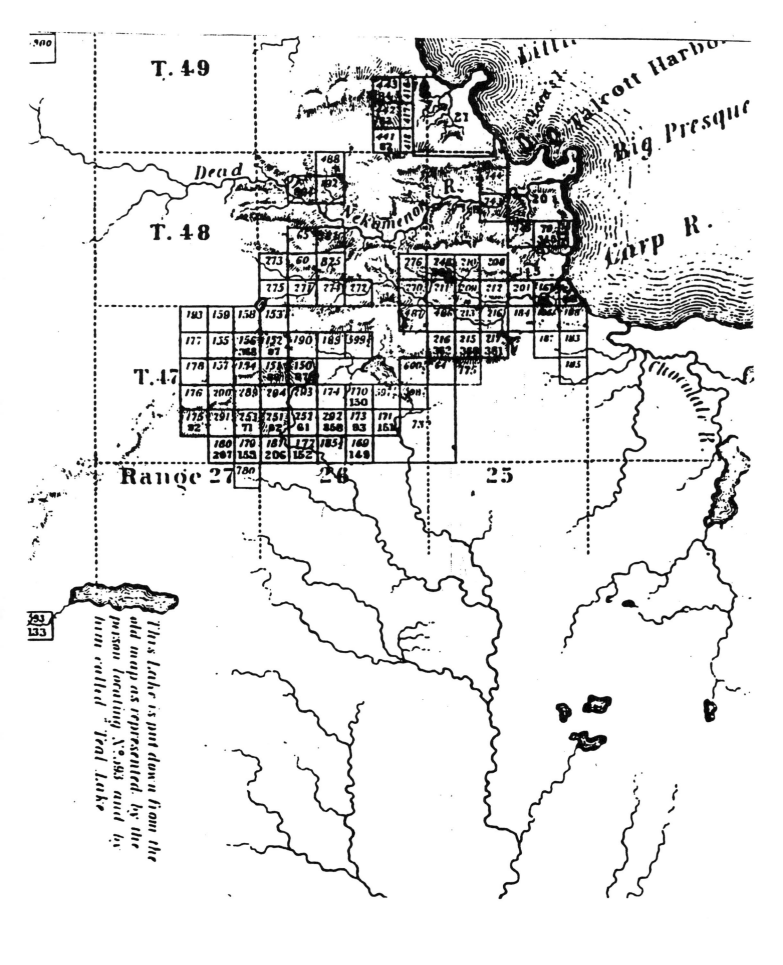

This Lake is put down from the old map as represented by the person locating No 593 and by him called 'Teal Lake'.

The final purchase[44] described 630.28 acres out of the 640 acres in Section 1, Township 47 North Range 27 West. This is 12 miles north and about 6-7 miles east of the location shown on these maps.

After the long walk to the iron mountain, Philo Everett had some doubts as to the value of the discovery. Family tradition says that he thought the mountain would supply all the iron the world would need for all time.[45] This and the Indians' skepticism about how the ore could be gotten out kept him from making the most of his opportunities, as he stated in his continuing Recollections[46]:

I did not lay a permit on what is now known as the Cleveland, believing, as Louis said, we had all the ore we could ever get away, of the very best quality and nearer the lake, preferring to let some other party take it and help to open the country. I had only seen this kind of rock ore once before. That was twenty miles from Black River in Oneida County, New York, between that place and Lake Champlain. That ore was precisely the same as the specular ore of Lake Superior. At Copper Harbor I met Professor Shepard, of New Haven, Connecticut, and I showed him the iron. He said it was as fine ore as he had ever seen, but thought it nearly worthless as it was so far away; it would be like lifting a weight at the end of a ten foot pole. But when I parted with him in the fall he said he had thought much about that iron, and believed I had better take care of it; the time might come when it would be worth something.

**PLATE 13**
Detail of Plate 2,
showing Lac Vieux Desert trail.

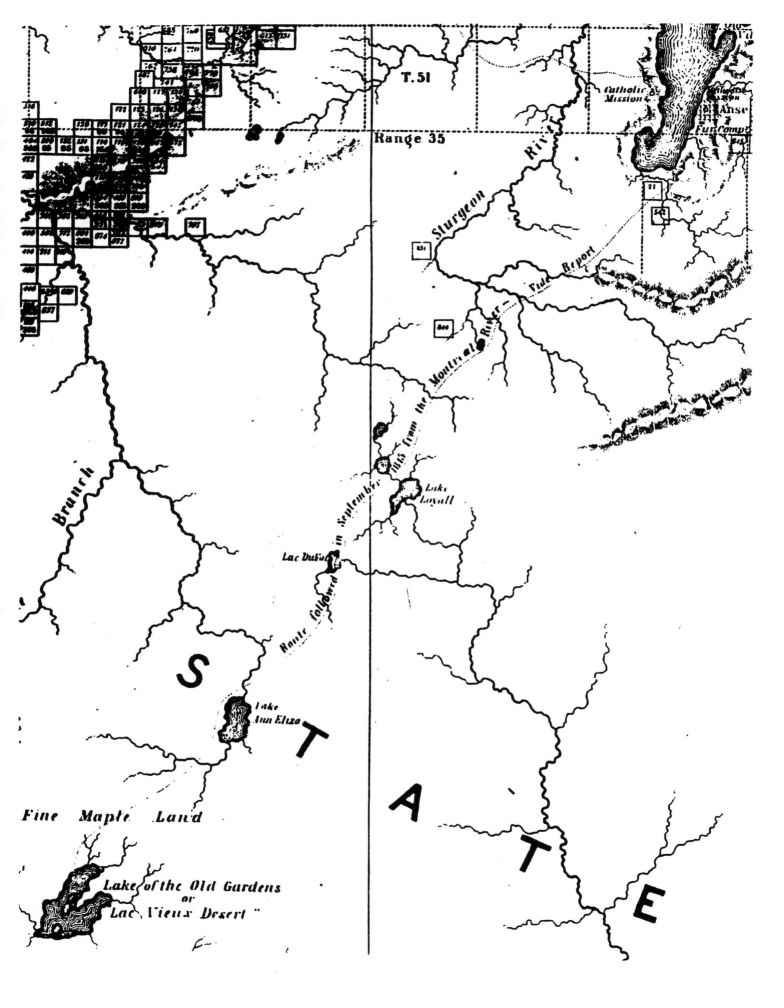

T. 51

Range 35

Catholic Mission

St. Anse
Fur Comp.

Sturgeon River

Tide Report

from the Montreal River

Route followed in September 1845

Lake Loyall

Lac Dufau

Branch

S

Lake Ann Eliza

Fine Maple Land

T

A

T

E

Lake of the Old Gardens
or
Lac Vieux Desert

We made explorations in various directions, locating several permits about Houghton,[47] but never paid any attention to them afterwards, turning our whole attention to the iron.


## The Side Trip

The main concern was now to return before the fall storms and winter. However, there was one more detour (See Plate 13) to L'Anse and Lac Vieux Desert, as Everett continued[48]:

"Louis had seen several pieces of copper in the hands of Vieux Desert Indians, but could not ascertain where they got them. Thinking there must be copper in that vicinity, I determined to go there and settle the matter, if possible, and left Copper Harbor with that intent. We ran ashore at the Catholic Mission at L'Anse. Several Indians met us at the landing and all knew Louis. It was soon noised about that Louis had come, and I believe all the squaws, old and young, came rushing down to meet him, joy beaming all over their faces. They all seemed to be meeting a loved father. They all bade him welcome with a warm kiss; it was really a pleasant sight. It was Saturday and we intended to stay Sunday there. We pitched our tent and moved our things into it. Several of the Indians came and seated themselves in the tent. Among them was a very old Indian but in full vigor, and about as homely a face as could well be conceived. I saw he had his eyes on our little vinegar keg most of the time. I asked Louis in a whisper what the old fellow thought was in it. He replied: "Brandy; can't you give him some?" "Yes, yes," was the reply. I bought that vinegar in Detroit, the very strongest to be got; no one could use it without watering it. The color was precisely like brandy. Louis took a cup, drew out a little, and handed it to the old Indian. He took one swallow. I was frightened, for I thought I had been the cause of killing the old fellow. His face was about as ugly as it could be, naturally, but such a contortion I never saw before. After a time he began to get his breath, and when the other Indians saw he was not dead they set up such a boisterous laugh as I never heard from Indians before. I expected the old fellow would be fighting mad, but when he got his breath, so he could, he seemed to enjoy the joke. Louis asked him if he would have some more brandy; the answer was quick and sharp -- "Kahn win" (No!).

I attended the mission church with Louis on Sunday; the services were conducted by Bishop Baraga, the singing was by Indian women and was very good.

Monday morning we again started out for the forest. The rocks southwest of L'Anse, beyond the Sturgeon, have a very singular appearance, of a reddish cast and streaked with white quartz, but we found no mineral.[49] When we arrived within half a day's march of Vieux Desert it was storming, half snow and half rain, making the bushes wet, so we camped on the trail about noon, pitched our tent and built a big fire to dry ourselves as well as to get our dinner.

After dinner Louis said he would go and get some ducks. In the afternoon several Indians came from the payment at La Pointe[50] and camped near us. An Indian never builds much of a fire, but they like a fire as well as a white man. One Indian, with a wife and boy, pitched his blanket tent opposite our fire, the others near by. A little before dark Louis came with twenty two ducks and two geese -- quite a load. I told Louis to give the geese to the Indians, as the ducks were as much as we could eat before they would spoil. He gave one to the family opposite

58

our fire and the woman dressed it, out it up and put into their camp kettle, boiled it a short time, and then the two, with the boy, ate the whole goose. Our ducks were so fat we could not eat them without roasting out the fat before the fire. In the middle of the night I heard a groaning and vomiting; it was the Indian opposite our fire. I asked Louis what was the matter. His answer was: 'Do' know, Indian very sick, guess he going to die.' It struck me it was raw goose. I said to Louis, 'Have his wife make some very strong coffee out of the corn, and give him one swallow every five or ten minutes.' I got up and watched to see if she did it right. She made it very nicely and commenced giving it to him as directed. In a few minutes the vomiting ceased, and in about half an hour his groans ceased also. In the morning I enquired of Louis how the Indian was. 'Going to get well," was the reply; "they say you are a great medicine man.' About noon the woman packed up and took the whole camp equipment on her back and left for home, the Indian following. As the weather had cleared up and the bushes began to dry, we packed up and moved also. This was the 20th of September.[51]

The Indians make great use of the wild rice growing in those lakes, sometimes in ten feet of water. It looks like a rye field, and the kernel is like rye, but only about two thirds the size. It is gathered mostly by the women, one pushing the canoe while the other bends the rice over the side of the canoe with a crooked stick similar to a sickle, and strikes it with a sharp knife, and the heads drop into the canoe. The ducks and geese stay here in large numbers until the ice drives them away. We found many pieces of copper in the hands of the Indians, but they seemed to be handed down from one to the other until no one could tell where they came from. These Indians were a filthy, dirty set of people. Louis called them 'wild Indians.'[52] I saw no sign that they ever washed their clothes, and don't believe they ever did. Dirty clothes was

59

not all they had on, as we could testify.  When we returned to Lake Superior our flannel had to have a thorough boiling.

The season of navigation on Lake Superior was now drawing to a close[53], and I turned my face towards home.


## The Letter Report from Jackson

On his return Philo Everett's most immediate report was to his brother-in-law, Capt. Gilbert D. Johnson.  It is quoted in full:

"Jackson, Nov. 10th '45


Dear Brother

Since I have returned from Lake Superior, Charles [Charles Johnson] tells me he promised to let you know all about my excursion there, which he wished me to do, so I will undertake to.

I left here the 23rd of June [as corrected in testimony], last, and returned the 24th of Oct.  I had some talks about going up there last winter, but did not think seriously of going until a short time before I left.  It was with a good deal of trouble that I could get anyone to join me in the enterprise but at last I accomplished it by forming a company of 13 [See list above].  No one can make a location in the mineral district without a permit from the Secretary of War.  We had seven permits and I was appointed treasurer and agent to explore and make locations..

I took 4 men [3 of them named above] from Jackson, and hired me a guide at Lake Superior; bought me a boat and coasted up the lake to Copper Harbor, which is over 300 miles from the Sault Ste. Marie.  There are no white men on that lake, but those who go there for mining purposes. I was most of the time with the Indians, and those of the wildest nature.

We incurred much danger and hardship. The lake is one of the most boisterous in the world. I have seen it when our sails would not flop, and in 15 minutes, blowing a gale, and the seas, in a few moments more, running as high as a house; and that is what makes it so dangerous for small boats to navigate. There are many bays to cross and some places the rocks are perpendicular for many miles and no landing at all. If a small boat is caught here in one of those common, but severe squalls, it must be lost or ride out the gale. We have often been wet for days together. When we left the shore to explore, we took one blanket each, and what flour and pork we could carry, and we were obliged to go ahead rain or shine, for our provisions were stinted for so many days, and we found ourselves short many times. I passed, one day, 14 swamps, and we could scarcely ever cross a swamp with dry feet, and at night, lay on the ground.

When we were coasting, sometimes we had land for our beds, at others, gravel or cobblestones, and sometimes the soft side of a rock. Once, I remember, we lay on the rocks near the shore and in the night the wind blew the water up the rocks to us, and our beds of rock, as well as our backs, were found in the morning to be in the water.

We found many curiosities, many good agates, and we made several locations; one we called Iron at the time. It is a mountain, 150 feet high, of solid ore, and looks as bright as a bar of iron just broken; but since I have got home, it has been smelted and produces iron and something else, some say gold and some say gold and copper. I have now a breast-pin on, and the best of judges cannot tell the difference from the best of gold; at all events, it is creating a great excitement here and in Detroit. What it will amount to I am not able to say. I have 200 shares and

I think it will get me out of debt. Shares are held at $50.00 but none offered at that. I think I could sell at $25 part cash but I am not anxious to sell until it is thoroughly tested. If there is gold in it I can make a handsome property out of it.

We have other locations of copper. All locations are one mile square. We shall send a company of men in the spring to make other locations. Our half-breed Indian is still in our employ with lots of other Indians, this winter. I think there is no doubt but that we shall have one location of lead and silver in the Spring. We had not time to survey it out this fall. Our company is called the Jackson Mining Company. We have had several letters from the broker in Wall Street, New York, applying for shares in our company. One man in New York owns 25 shares in the Company. I send you a plat of our company. I have promised Charles 5 shares. The copper fever rages here more than any other.

It has been very healthy here this summer. All of our friends have enjoyed good health this summer with but few exceptions.

I wish if I have any drills or any wheel barrows at Whiteport[54] you would knock off the irons from the barrows and send them to me the first boats in the spring. I want them sent no further than Detroit to E. B. Hastings or Matthew Bissell. But if you can sell them for half what they are worth do so in preference to sending them west. All that I have paid to you in relation to coffee, etc. keep to you. I wish you would see Paulling and ask him to give you a statement of our indebtedness to Paulling & Co. and if he will take securities in different payments for the amount due them. We can sell our farm I think by giving some time. I think we can get $100 down and the rest secured. If they will consent, we will sell & give them all the first payments sufficient to pay them up. This is all that is in our power to do at present. We are anxious to close up in some shape. I have written them several times and got no answer.

If they should feel dispose to collect by law it will only oblige them to pay cost which I am not disposed to have them lose. I want they should have their due without paying it to lawyers and I offer to do all that I can. The offer will secure them their pay in a short time sooner and quicker than I could pay.

P.M.Everett"

The last two paragraphs of the letter were omitted from the version previously published by family members (CRE p.433 and Mrs. Everett) and have been taken from a photostat of the original, copies of which are in the Michigan Iron Industry Museum.

Note that Everett was already planning the mining or quarrying of the iron deposit. He was also recalling his prior experience as a builder of canals and kilns.

## The Reorganizations of The Jackson Mining Company

At this point, it might be well to review the economics of The Jackson Mining Company. Although everyone talked about shareholders (at least in hindsight), it was not at first a corporation, but an association of thirteen members. Mr. Everett testified that they contributed $50 each. Prior to the date of the above letter (Nov. 10, 1845) each received 200 shares, i.e., at a cost of 25 cents each. These were assessable but 500 shares were authorized to be sold as paid in full and unassessable, and these were apparently sold by 1846. If each of the shares was now valued at up to $50, then many people thought that they had a valuable find. The interest in "Wall Street" and the sale of shares in New York, all developing within two weeks from

Everett's return, shows how inflammable the market in Lake Superior mining stock had become, although up to then it was based entirely on copper and, as his letter said, remained primarily a "copper fever".

## Events of 1846

Despite the apparent sale of 500 more shares at advanced prices, the members were cash poor and, as Everett had anticipated, had no resources to exploit the find in the following year, but they did send two members up to correct the location of the claim and to bring down a little more ore. Everett's Recollections[55] describe this as follows:

> In the winter of 1845 and '46 I learned that the township lines about Teal lake had been run in the summer of 1844.[56] Towards spring two of our company, Col. Berry and Kirkland (sic) [President and Secretary of the Company], offered to go up the lake in the spring, if the company would pay their expenses, hunt up the town lines, and go to the mineral office and locate Teal lake correctly; their offer was accepted, and they were instructed, if they met with a party they had confidence in, who would promise to keep open the country, they might show them the iron two miles beyond ours and let them take possession of it.

Everett does not explain why he did not return to Lake Superior in 1846 or 1847. He followed and describes the events of those years so closely that some writers assumed that he was more intimately involved and did return during those years. He had, of course, a business and a family to attend to in Jackson. He may have been financially pressed. He did not wish to sell his shares as yet, but did wish to sell a farm and to settle a debt, as mentioned in his 1845 letter. His family was growing -- a last son was born late in 1847.

Everett then summarizes one result of the Berry trip[57]:

"Berry and Kirkland (sic) found the township corner only a few rods from the southeast corner of Teal lake. Our boundary started at the southeast corner of Teal lake. Three quarters of a mile southwest[58] was the outcrop of ore, making that the center of the section; this was now known as the northeast corner of town 47 north, range 27 west, and section 1. Then they went to Copper Harbor and had Teal lake correctly placed on the mineral map, more than twelve miles from where it was first laid down."[59]

Berry tells the story with (naturally) more emphasis on his role, as follows[60]:

In the spring of 1846, another expedition was fitted out, consisting of F. W. Kirtland, E. S. Rockwell, W. H. Monroe [these two from the previous group] and myself; the object being to make further examinations of the iron and to use the remaining permits, to enter other mineral land. ... After spending twelve days in the woods, exploring the surrounding country, including what was afterwards known as the Cleveland and building what we called a house, we returned to the mouth of the Carp with 300 pounds of ore on our backs. We then divided; one party was left to keep possession of the location, another went farther up the Lake to use the remaining permits, while I returned to the Sault with the ore. It was my intention at this time to use another permit [This must have been their 7th and last permit, which in fact was never used.] on the Cleveland location, but on arriving at the Sault I met Dr. Cassels, of Cleveland, agent of a Cleveland company, and having arranged with him that his company should pay a portion of the expense of keeping possession, making roads, etc., I discovered to him the whereabouts of the Cleveland location,

He took my canoe, visited the location, and secured it by a permit.[61]

The more important event -- at least as far as future litigation was concerned -- occurred at Dead River (River Du Mort) when they gave Marji Gesick a certificate for "twelve undivided thirty-one hundredths part of the interest of said Jackson Mining Company in said location No. 593." The following is a copy of the certificate as delivered, from one of the Court's decisions[62]:

"River Du Mort, Lake Superior,
May 30, 1846

This may certify that in consideration of the services rendered by Marji Gesick, a Chippeway Indian, in hunting ores of location No. 593 of the Jackson Mining Company, that he is entitled to twelve undivided thirty-one hundredths parts of the interest of said mining company in said location No. 593.

A. V. Berry, Pres.
F. W. Kirtland, Secy."

There may have been some difficulty in interpreting this certificate. However, it seems to have been intended to give Marji Gesick the equivalent of twelve shares. There was discussion of this after Berry and Kirtland returned. While there were objections, Everett indicated that some time, perhaps that winter, the members agreed "they had better acknowledge it, accept it, and they did so." The trouble was that all the unassessable stock had been issued and there was none left. As late as 1848, shortly after the Company had become incorporated, there was still no solution to the problem, as indicated by the following from the records of the Company[63]:

66

June 7th, 1848. Inquiries were made by James Ganson concerning the unassessable stock, and if eighteen shares had been reserved for certain Indians near our works; it was shown by Mr. Rockwell that said Indian or Indians were promised eighteen shares for services rendered at the time the iron mountain was located, and by Kirtland and others that written certificates of the same (or 18-3100 of the iron location, though of this there seemed different opinions) had been given to Marji Gesick and other Indians, and signed by himself and Colonel Berry in the spring of 1846. As no entries of the transaction had ever been made on the certificate book of the company nor any official report of the said transaction had been made, said number of shares had not been reserved from the five hundred shares of unassessable stock, and as there were no more in the possession of the company, the following was offered by Mr. Foote, viz: 'Whereas, It appears satisfactorily to this company that 18 unassessable shares of stock in this company were issued to certain Indians, *Resolved,* That the committee heretofore appointed to report the plan and condition by which the charter of the Jackson Mining Company passed by the legislature of the state of Michigan, and approved April 3rd, 1848, can be adopted by this company, be instructed to report some provision by which certificates of stock may be issued to said Indians or their assigns.' Which resolution was passed unanimously by a vote of ayes and nays.

Everett's Recollections[64] round out the report of the 1846 activities:

Soon after they [Berry et al.] returned to Jackson, I went with several persons of the company to an old forge called Hodunk, a few miles north of Coldwater, Michigan, taking with me some of our iron ore. This forge was run on bog ore, Mr. Olds being the forge man and William

Lemm the helper. They took our ore and made a bar or iron from it in our presence. On returning home to Jackson I took the bar of iron to John A. Bailey, then a resident of Jackson. He converted it into steel and made a knife blade on one end, and ground it thin like a razor. It would cut hickory bark without turning the edge. It was as fine stuff as I ever saw in any instrument.

It will be noted that this was the second report of making iron from the ore. In still another version Berry says (in the Transcript and in the report quoted above) that on his return he made two attempts to smelt it in a cupola furnace and failed. Some of the ore, he continues, was then taken to Mr. Olds of Cucush Prairie, who succeeded in making a fine bar of iron from it in a blacksmith's fire, and it was stated that this was the first iron ever made from Lake Superior ore (ignoring the breast-pin described in Everett's letter quoted above). Berry called "that one bar of iron ... the great enter[ing] wedge for the whole iron district". (Tr. 102)

## Everett in Jackson from Late 1845 until Spring of 1848

Despite the interest with which Everett followed the affairs of the Jackson Mining Company and the detail in which he reported the events of 1846 and 1847, he remained in Jackson all that time. One writer speculates that he was ill, even that he had been sickly throughout the '40s. His later memoirs protest that he was in good health all this time.

Contemporary records bear out the belief that it was the continuing care of his business, which he did not dispose of until 1850, and family matters that were his concern for this period of over two years. More specifically, as mentioned above, his youngest son, William, was born on Christmas Day in 1847 and christened with his parents as sponsors on April 22, 1848 in Jackson.

## Developing the Mine - 1847 to 1850

The previous narrative anticipates what a problem it would be to exploit the mine. Both Louis Nolan, as a practical man, and Professor Shepard thought that getting the iron out to the lake was probably uneconomic. Apparently they thought that getting the ore away from the Cleveland location two miles farther on was really impossible.

Unlike the pure nuggets of native copper being shipped from Keweenaw, the iron ore was heavy, of lesser value and needed to be converted to iron. This first solution was to build a forge heated with local charcoal and with a hammer run by water power. Then blooms or ingots of iron of increased value could be delivered to the shore of Lake Superior. It will be remembered that there were two steamboats and only a few schooners on Lake Superior and nothing else was available except canoes, bateaux and Mackinaw boats.

The iron was not to be cast from a blast furnace, as was done in the larger centers at that time. Instead the centuries-old method was followed -- heating iron ore with limestone and charcoal in open forges until blooms were made. Water power was needed to operate heavy hammers that produced ingots of wrought iron. Prodigious quantities of charcoal would also be needed, but Coal wood could be cut and burned in stacks close to the forge. Later, and only when there was a passable road, it was burned in great kilns or ovens, remains of which still stand near the mouth of Carp River. But shipping the ore itself out to the lake was out of the question.

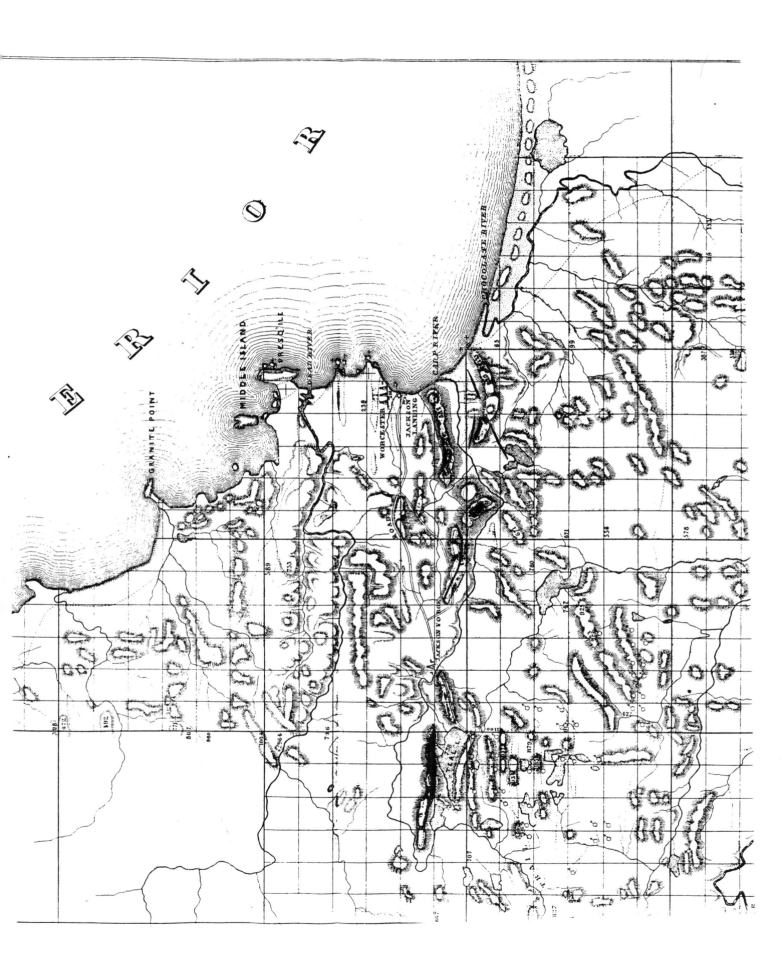

The Everett Recollections[65] condense the developments followed (being ambiguous about his whereabouts up to "the spring of 1848"):

Having settled the fact that the iron was of the very best quality, we at once set about preparing to build a small forge to bring out the iron and make it known. Men were hired and sent up to build a dwelling[66] and warehouse at the landing, build a wagon road to the mine, and a dwelling at the mine.

Along this route, they found a site for a forge, where there was a small waterfall. There they planned to settle and build cabins. Meanwhile the forge had to be designed and the bellows, heavy hammers and other equipment were ordered in Jackson.

Everett continues:

I found it difficult to hire men, because they were afraid of suffering with the cold, believing they would freeze to death in that cold region. We hired a man from Pennsylvania, a forge builder, to go up and put up a forge on Carp river, two and a half miles east of the mine; his name was McNair.[67] We had him made postmaster, calling the office Carp River post office,

It was well into 1847 before much could be done at the site. As far as previous action went, McNair had been considered a "fanciful mechanic" and "they didn't do much that fall" of 1846, according to Kirtland's later testimony.[68] Although an assessment of February 11, 1846 was supposed to have raised a considerable amount, money was a problem. Everett had sold his farm, settled his old debt and now could loan the Company $700. This was a little more than the Company's original capital and seem to have enabled the company to send out a new party.

71

**PLATE 15**

View near Carp River, Lake Superior

(Marquette, ca. 1851)

(The new corporation reported at the end of 1848 that, with another assessment, it repaid Mr. Everett and "that there had been received into the treasury up to Dec. 19, the sum of $3,148.28, and paid out ...$2708.43".)[69]

In July 1847 a competent party left the Sault on the steamer *Independence* and landed at Carp River. They were led by S. T. Carr of the original party. The crew to operate the forge was headed by Aaron Olds and William Lemm, who had made the first bar, with Olds's brother-in-law Ariel N. Barney.[70]

They wrestled the heavy equipment over a mere trail to the forge. They then built two bloomeries and an earthen dam to operate them. A total of twenty-four men and two women wintered over in one large cabin. In mid-winter (February 10,1848) Olds and Barney made the first bloom and a bar from it.

In an article entitled "From a Forge a Town was Born" Ernest Rankin said:[71]

...Into the Winter of 1847/48, ...[the] workers first built a trail and then widened it to an extremely rough road over which they hauled the material for the forge as it was received from the vessels coasting up and down the Lake. ... their materials, tools and provisions were brought from the Sault to the mouth of the Carp River in the *Fur Trader*, 51 tons burden, one of the very few small schooners then operating on Lake Superior. It was not only difficult to get a ship to carry their freight, but, as the mouth of the Carp was open to the storms of the Lake, it was at times impossible for a ship's captain to discharge cargo at that point. [See Plate 15.] Frequently the schooners continued on to Copper Harbor and Ontonagon, hoping to enjoy better weather on the downward trip. Sometimes it was possible to unload up within the still waters of the "Dashing River," (later to be known as Dead River), its entrance being afforded some protection by Presque Isle, a short distance to the north. This added another three miles of road, for the most part through heavy sand, over which the freight had to be hauled. It was not only an isolated, rugged country for men to work in, but it required the hardiest of men to withstand the rigors of a 16-hour, day-after-day battle, to obtain a foothold in this frontier land.

Everett reports[72]:

The spring freshets in April took away part of the dam, and, on the opening of navigation, McNair came to Jackson. We settled with him and he returned to his home in Pennsylvania ... [Everett attended a Company meeting on June 6, 1848 and soon thereafter] I went up to repair the dam, taking my wife [Charles Johnson, and two sons] with me. [In old age Mary Huntoon Campbell told of coming to Marquette with Mrs Charles Johnson . Probably she was on this trip and they both returned with Mrs. Everett in the fall.][73] My wife returned to Jackson in the fall, but I [we] spent the winter.

Only a small amount of iron had been made before Everett took over. He first repaired the break in the dam and started making iron in September. Because of the makeshift nature of repairs to the dam and other "various causes", such as lack of ore and charcoal, by November he had made only ten tons and was able to get two tons to Detroit before navigation closed. He could, however, report in the last communication of the year, that the dam was finally repaired and the works were making 1800 pounds of blooms per day.

Everett's memoir adds:

The Forge was [re]started in the winter. In the summer of 1849 I built a sawmill and in the fall Charles Johnson [, the boys] and my self returned to Jackson.

Rankin describes the communication problems[74]:

The iron ...[was] transported in wagons, six horses to each, over the treacherous road to the mouth of the Carp to await the arrival of a downbound vessel. [It was also sent in the winter by sleigh, but that was not much easier as the snow would not hold up.]

The first post office of the Iron Region was established at the Jackson Forge Location, Carp River, on January 12, 1847. One of Everett's men, a William B. McNair, was appointed postmaster, which position he held until June 29, 1849 [1848]. Everett was then appointed, serving until April 2, 1851, when the office was discontinued. Little mail was handled after the close of the navigation season as Green Bay, Wisconsin, the nearest connection with the outside world, was some one hundred and fifty miles distant. Winter mail deliveries, consisting of a dozen or so letters, and a few newspapers to be shared by all until worn out, were [much later] made about every six weeks. While no mail service was ever authorized on the boats plying between the Sault and the Lake Superior mines all letters accumulating there during the navigation season were forwarded in care of the officer in charge of a vessel.

## Rival Claims in 1848 and 1849

There had been no interference with the Jackson claim before the summer of 1848. Then, however, the story that had emanated from Jackson brought new claimants and would-be claim jumpers. Everett tells[75] the further narrative of the Jackson mine and its rivals in detail:

Robert Graveraet, Samuel Moody and Mann arrived there before me [in 1848], and went on up to the mines, and, in the absence of the man in charge of the Cleveland location, burned the building the Cleveland company had built, and built one of their own, and gave out that they would shoot the man in charge if he attempted to return to take possession. After their building was completed, Graveraet returned below.

As Everett reports later, this attempt was disallowed. On the other hand, a friendly claimant next appeared. Everett's Recollections continue[76]:

A few days after I landed there, John Burt, with a company of surveyors, landed, Mr. Burt having a contract to survey out several townships of land west of the forge. He packed his supplies past the forge and the Cleveland to the "little location," as it was called. Believing the little location was worth taking care of, he put up a building there and made it his general depot for supplies, and from there supplied his surveyors west. When his surveying job was finished in the fall, Mr. Burt left two of his men in charge of the location to remain there all winter.

Soon after I arrived at the forge, I received the appointment of postmaster in place of McNair, resigned. Shortly after the arrival of John Burt, Mr. Foster, of the firm of Foster & Whitney,[77] came and wanted all the information I could give him about the iron. I found that our men had rambled over the country quite a long distance, as far as what is now known as the Republic mine. I gave the description of that location, as well as many others. He went and made thorough examination of the mine, and when he returned he told me it was a bigger thing than the Jackson mine. That was hard to believe, for we thought the Jackson could not be beaten.

Next came Alexander Sibley, of Rochester, New York. He came up the lake on the schooner Furtrader, sailed by Capt. Ripley. We had a contract with Capt. Ripley for our freighting and he usually called on his way up and down the lake, and promised to call on his way down for him. Mr. Sibley told me his business was to purchase the Jackson mine and property, if it suited him. I had a couple of horses saddled and rode out with him to the mine. He examined the bluff at the foot of the elevation very closely and evidently with much surprise. We then began to climb the hill; he looked closely at the iron outcrops as we went up. When about half way up he stopped, saying he had gone far enough. I remonstrated; told him there was a big bluff on the top of the hill where a hundred tons or more of ore had been thrown down by the frost. He replied it was of no use to him -- there was too much of it. 'You say,' said he, 'there are two more mines two or three miles further on, and here is ore enough to supply the United States for all time to come. This hill is one solid mass of the richest iron ore for half a mile and how much more no one knows; it will only be worth the digging.' We came back without making any further examination. I entertained him until the Furtrader returned,, and he went back to Rochester, went into the telegraph business, made a fortune and was called the Telegraph king of Rochester. He was a very intelligent, pleasant gentleman, but the iron was too much for him; he

would not touch it; if there had been less he would have
bought it out.

Next came Mr. Jones, my neighbor, of Jackson. He was
not a stockholder, but hearing so much said about the iron
mountain he determined to see it himself; he spent several
weeks with me. When he returned to Jackson he began buying
up the Jackson stock, and got enough to elect himself
president of the company. ...

Mr. Jones was, in fact, a principal holder of the
unassessable stock of the Company and spent much of the month of
June, 1848 contesting the mechanism for converting it into stock
of the new corporation. He had accomplished a satisfactory result
before leaving for Lake Superior.

## Another Attempt at Eviction

Everett continued:

In the winter of 1848-'49 Robert Graveraet went to
Mr. Clark, of Massachusetts, and represented to him that the
controlling stock of the Jackson Mining Company could be
bought very easily. Mr. Clark was not a capitalist but a
kind of speculator. He induced Mr. Fisher, a heavy cotton
manufacturer of Massachusetts, to furnish the money to buy
out the Jackson, or the controlling stock, and let him
manage the concern. A. R. Harlow had a small machine shop,
and he agreed to put that into the company and move on and
take charge of it as a member of the company. Now all was
ready by the opening navigation.

A month before the Harlow party could reach Marquette,
Joshua Hodgkins (who had married one of the Everetts' nieces)
came up from Jackson to work at the Forge. He tells how they
arrived on June 5 and he, with his wife and two sons, moved at
once to the Forge, where there were about 40 workmen. [78]

**Everett continued:**

Clark and Harlow came on to Detroit and purchased a large stock of supplies in the way of eatables, but no tools or teams were bought, as they were to take possession of the Jackson Company and all that belonged to it. They did not wait to see if they could get the stock, but shipped their supplies on to Marquette with Mr. Harlow and his family; Graveraet was also one of the company. Clark went to Jackson to get control of the stock, as he had been told it was an easy affair. Graveraet and Harlow with their supplies arrived at our landing in Marquette [July 1849] while I was at the forge, and took possession of our dwelling, expecting to get possession of the forge and all things thereunto belonging, together with the mine, in a few days. When I came down to the lake on business, I had to seek quarters with the Indians or sleep out of doors. [Kawbawgam ran a log "hotel" that was an improvement on the typical Indian bark shelters found in Indian Town just below the mouth of the Carp. See Plate 16.]

Clark went to Jackson from Detroit to purchase the controlling stock of the Jackson, but Mr. Jones learned what was going on and got the control of the stock in his own hands, so that when Clark came there he soon found he was foiled. He returned to Detroit and took the first steamer for Lake Superior. When he reached Marquette he found things decidedly flat; they were there with a large supply of provisions, but no tools or teams and no laborers, and were trespassers.

PLATE 16

Wigwam of Jacques LePique, near Marquette

They held a council, and determined to go ahead and build a forge of their own at the shore; they built a log building for their supplies, and went to Dead river, took down a log dwelling that was there, brought it to Marquette, put it up near where the Cleveland dock now stands, and vacated our building. They were in want of men, so Clark left for Milwaukee, hired forty men, French and Germans, and put them on a sail vessel for Marquette. At the Sault he stopped off and was taken with the cholera and died in a few days.

Graveraet and Harlow set the men cutting [char]coal wood. It was curious to see them chop; never having used an ax, they clinched the handle of the ax tight and held on, not slipping either hand. In cutting down a tree, they chopped all around the tree, and when weakened so as to fall, it went the way it leaned. There were forty men in a huddle, and when the tree started a yell was given so that every one could run for his life. To chop off a log they stood on the ground, if the log was ever so large. But the little machine shop went up, and brick making and building went on.

[In a humorous fashion, Peter White tells how this same party of greenhorns tried to build a dock, only to find that it had completely disappeared when the Lake began to blow.]

About this time the Cleveland company represented to the commissioner of the land office at Washington in what manner Graveraet, Moody and Mann had taken and retained possession of their mine, and the commissioner awarded the mine to the Cleveland company.

PLATE 17
Portion of map,
*Plan of the Village of Marquette*

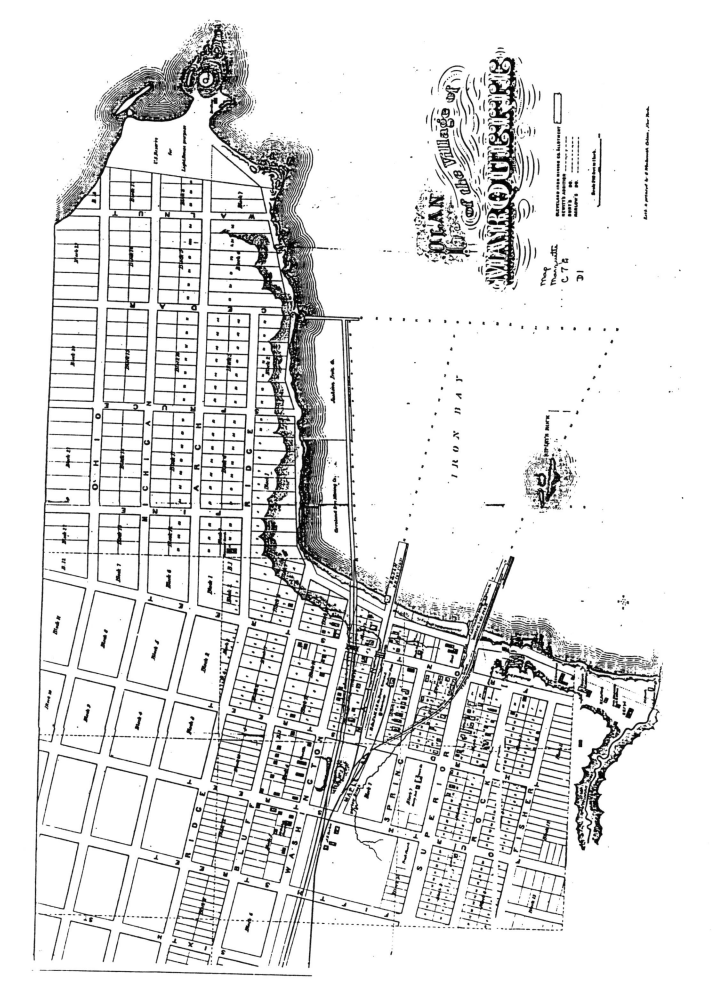

When Moody and Mann heard of the result they packed up and joined the Marquette company at the lake. When the land about Marquette came into market John Burt appeared at the land office at the Sault to enter his 'little location', but Graveraet opposed him so sharply that Burt settled the matter by giving him half of the mining right.[79]

## Everett Turns to a Different Venture

As noted above, the control of the Jackson Mining Company, after some unfriendly "raids", had passed to a Jackson neighbor. Nevertheless, Everett was concerned that, without the Eastern money exemplified by Mr. Sibley, the original group (and he in particular) would find the mining operation too large to handle. A more feasible venture appeared to him after Mr. Jones took charge of the Jackson property in the latter part of the summer of 1849 as he and Charles Johnson were returning to Jackson, as described above.

The Jackson Mining Company, according to the President's Report of December 27, 1848, had already secured "a deed of 51,50 acres of land at the lake embracing the harbor". This was apparently included in the claim that Farrand and Western on February 1, 1849 entered to 103 1/2 acres on the lake shore in what is now South Marquette running south from about Fisher Street. (See Plates 17 and 18.) As for himself, Everett reports: "While at the Sault we entered what is now known as the thirty six acre plat, in the city of Marquette, but yielded one half interest to the Cleveland company. That entry covered the land where the M. H. & O. merchandise dock [marked "B.de N. & M. R.R. CO'S PIER" on Plate 18], the great ore dock, and the Iron Bay furnace now [in 1887] stand."

**PLATE 18**
Detail of Plate 17,
showing the "36 Acre Plat"
(aka. Cleveland Iron Mining Company allotment.)

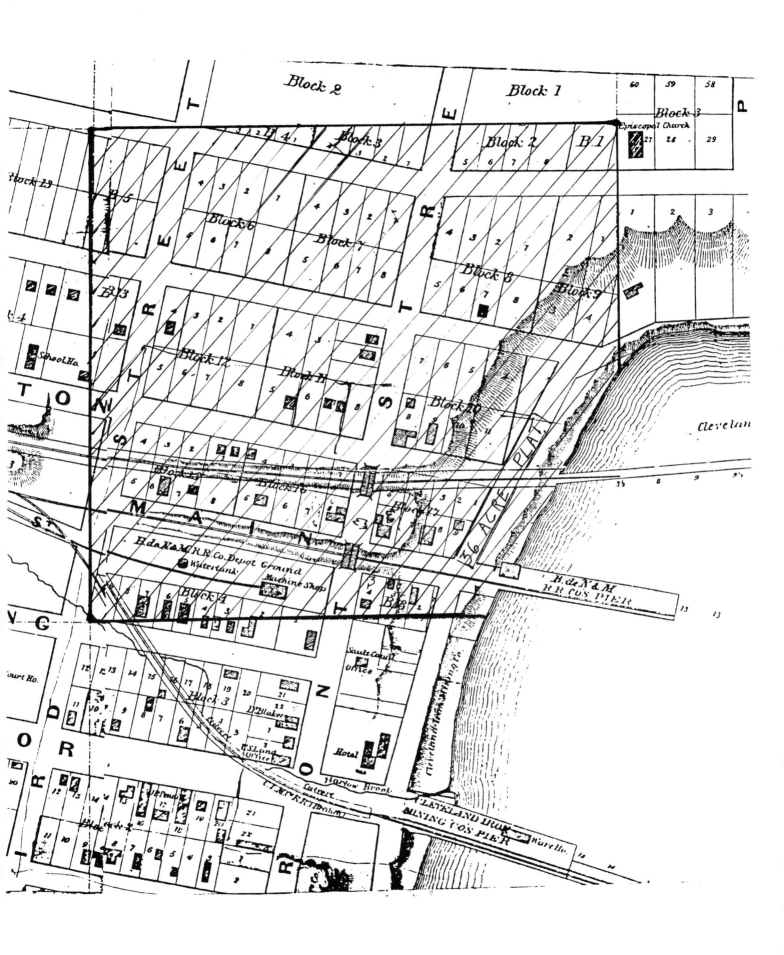

Harlow's group (Marquette Iron Co.) was fearful "that the Jackson Company had bought all the land fronting on Marquette Bay". Accordingly he and Fisher bought several lots at the Picnic Rocks and around the mouth of Dead River.[80]

The 36 acre plat (See Plate 19 for the official version), which laid the foundation for their subsequent fortunes, comprised the heart of the Marquette business district and was known as the 'Cleveland Iron Mining Co. Allotment'. Subsequent Additions encircled it.

In informal "partnership" with the Cleveland group, Everett and Johnson could proceed for many years in building houses and larger structures. In August 1854 formal partitions were filed, giving reservations to John Burt, the Cleveland Co. and the Plank and Rail Road Co. to cross the 36 acre plat to the shore and the docks being constructed. The lots were also formally divided between the Cleveland Co., Charles Johnson (acting for himself and Everett) and others.

Although they had granted rights of way for the roads later laid out by Burt and other friendly interests, some of the rival companies accused them of trying to monopolize the shore of Iron Bay and freeze out the others.

PLATE 19
The 36 ACRE PLAT
as recorded in 1855

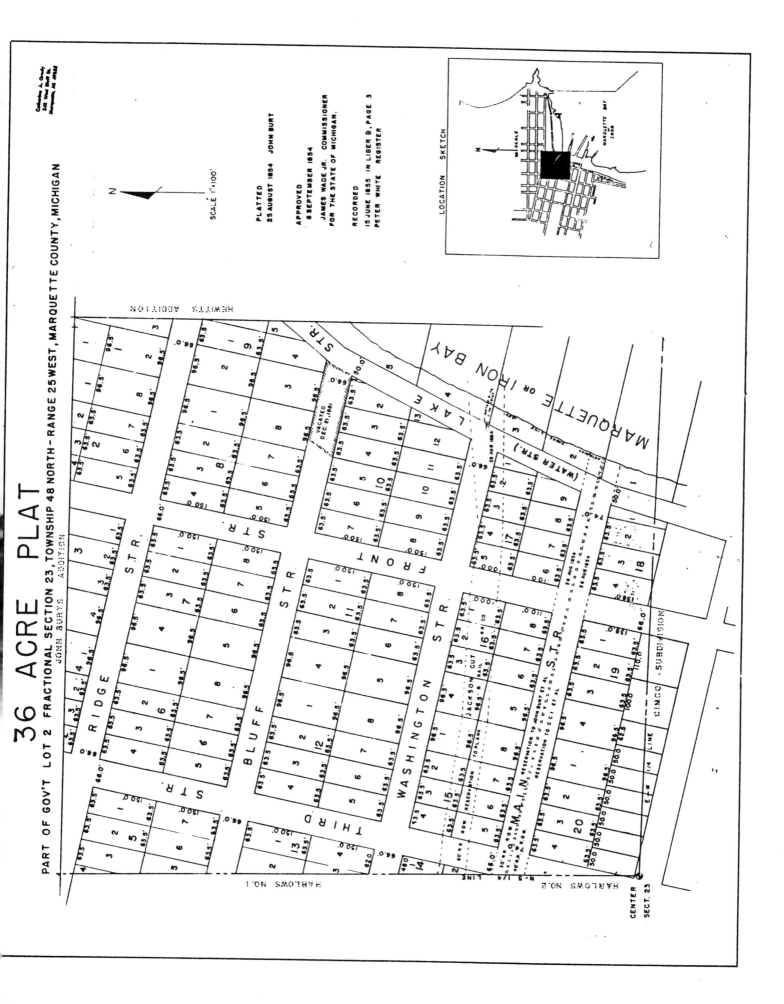

## The Move to Marquette in 1850

Everett's Recollections continue[81]:

I spent the winter of 1849 and '50 in Jackson, and in the summer of 1850 disposed of my warehouse and some other property, together with my Jackson mining company interest,[82] and in the fall prepared to move my family to Marquette [See below]. I left my dwelling to be rented, and most of my furniture to be sold, and Charles Johnson and myself and families left for Lake Superior. We boarded with A. M. Barney, who kept a sort of hotel on the ground where the Northwestern hotel afterwards stood. We soon finished off a small one story dwelling of two rooms and four small bed rooms on the ground floor, where the Burt block was afterwards built. [The dwelling must have been very "soon finished", as the recollections quoted below speak as though they moved in to it at once.]

C. R. Everett now tells this part[83]:

In 1850 the Everett family, consisting of the father, mother, daughter Emma Eugenia, born at Kingston, New York, daughter Ellen Mehitable, and sons Edward Philo and Charles Marshall, all born at Jackson, Michigan, moved to Marquette, and occupied a house built by Mr. Everett and his brother-in-law Charles Johnson, which stood [on Lot 6, Block 17] on the east side of Front Street north of where the Duluth, South Shore and Atlantic Railway was to cross the street. This house was later moved to the north side of Main Street between Front and Third Streets [Lot 6, Block 16 on Plate 19]. The house was described at the time of its being torn down, as a wooden structure of four small rooms and a lean-to. The youngest daughter, Katherine Eliza Johnson Everett was born in this house.

PLATE 20

The Burt Block and Front Street, ca. 1865

*(Courtesy of the Marquette County Historical Society)*

88

At this point, C. R. Everett's article quotes Mrs. Ellen Everett Robbins (Aunt Ella), one of the daughters, on the first winter in the little house. (The slightly longer quotation that follows is taken from her original letter to him of January 3, 1949.[84]):

Yes, of course, we had a very simple house on Front Street. That whole block [Block 17] & both sides of it between the two R.R. cuts belonged to father and there was our first little home, not much more than a shanty but very comfortable for pioneers. In that little house we had one large living room, which had to be Kitchen, dining room & parlor, and three bed-rooms for we were quite a family to be taken to such a wilderness -- Emma, Edward, Charles and myself -- & Kittie a little baby. Our living room opened into Uncle Charles' part and they had a big living room & two bedrooms. Mary [Huntoon] Campbell was adopted daughter of Uncle & Aunt Mary (They had no children.) [This was probably not a formal adoption, as Mary continued to use the Huntoon name and did not call the Johnsons her parents.]

Everett's younger son, Charles M. Everett, also described the house in later days [85]:

The room in this house would, in the present day of house building, be considered insignificant, but the building proved to be a veritable mansion at that time, and most hospitable, for in 1851 the following persons were sheltered that at one and the same time, without great discomfort, and with none of the social discords sometimes apparent under such conditions:- Mr. and Mrs. John Burt, Hiram A. Burt, and Alvin C. Burt, fourteen persons,[86] nine adults and the rest children, all living in the same house at once.

'How did so many people get along so well in so
small a house?', he was asked. 'Very comfortably and well.
All of the cooking and washing was done in the lean-to, or
out of doors entirely, while the living was done with a room
to each family. Not much room was required during the
daytime as the men folks were at work, the children down on
the beach or roaming the hills, while the women folks could
get along in the house very nicely.'

'But what did you do when night came on?'

'Oh, we packed ourselves away. We boys could
sleep on a crack in the floor, and then, too, outside of
bedsteads, bureaus and chairs, with a table or two, we had
no superfluous furniture. Our sideboards were stumps around
the house; for sofas we had moss-covered rocks or the sand
heaps, as we might elect, while for bath rooms we had the
whole of Lake Superior.'

'How did you boys amuse yourselves?'

'Day times we chased chipmunks, pulled ground
hemlock to sleep on, waded across the Carp river, fished,
hunted, and had fun generally. In the evening we generally
had a tussle for the soft spots in the floor to sleep, and
how we did sleep !'

Aunt Ella continued in her letter to C. R. Everett:

Of course, Father went to that country with a ship
load of men for all kinds of work besides mining [in 1848]
two years before he took the family - so that house was
ready for us when we landed. Uncle & Aunt & Mary [Huntoon]
were with us. Mary & [your] Aunt Emma [at the age of 11]
were quite grown up at that time and should have been left
in school at Jackson.

The C. R. Everett narrative[87] records those first days in Marquette from an article written by Mrs. Everett in 1879. Although it is rich in detail, it is hard to imagine the perils and hardships that she undertook, especially in her pregnant condition:

She wrote very entertainingly and interestingly of the early life in Marquette, and one article written by her, referring to happenings on December 15th, 1850 and January 1st, 1851...[is here given] to show how close to what might be called disaster the little settlement came. (Mrs. Everett's article was also published (posthumously) in 1921, Vol.V. Mich. Hist Mag. pp.569-573. It follows:)

It often brightens the present to look back upon the past. In 1850 we looked upon a small number of houses, scattered here and there among the pines, mostly built of logs, and one small store, from whence all the necessaries to sustain the little community were distributed -- mostly in small allowances -- so that none would fare better than his neighbors, and all depended upon the good or ill success of the forge that was to commence making bloom iron, as soon as the ore could be brought by the sleighs from the mine. THAT was Marquette. In the autumn the little steamer Napoleon was chartered to bring supplies and, as usual, attempted to do more than she was able to accomplish. Being overladen with freight for upper lake ports she <u>passed by</u> Marquette, hoping to be able to call here with a full load after it would be too late to go further up the lake. There was then no mail communication between this place and the outer world except by boat, consequently there was fear that we would be left without food for man and beast. November came with its storms and snows and <u>still no boat</u> came; winter seemed closing in upon us as the days went by, and all eyes and hearts ached in vainly looking so long and

91

anxiously for the Scootie-nobbie-quon (as the Indians called the steamboat). The first thing in the morning, and the last at night, was to cast a long look on the sea of waters, and turn away with a sickening fear that there was little hope that relief would ever reach us, but with every prospect that want and famine would be in our midst. All business was at a standstill; it was no use to go on with the work, for everything depended upon arrival of supplies. December 1st dawned cold and stormy; all hope seemed gone, and there was nothing now to do but look over the remaining stores to see how long they could possibly be made to last, and how much we could divide with our neighbors. At last it was decided to kill the horses, and divide the coarse feed left among the most needy families of women and children, and send the men away through the wilderness. December 13th and 14th dawned mild and hazy, and hope revived a little. The morning of the 15th I fancied I saw a faint smoke now and then through the haze, and we watched it long and anxiously. My breakfast remained untasted, and when fully convinced it was the smoke of a steamer, I told two young ladies they might go out and shout PROPELLER! PROPELLER!! as loud as they pleased, and then every home sent its inmates to swell the cry. Men shouted and swung the remnants of hats, women tore off their aprons and waved them, and the little feet that were bare for the want of the shoes that were on that boat, danced out in the cold, and their owners shouted too. One man whose fine span of horses had given out the day before for want of food, and one of which had been put in an old shanty to die, exclaimed: `Now, if old Bill is not dead I can save him.' There is probably no one living in this place now, and who did not witness this scene, who can imagine the feelings of the crowd that greeted that boat, and made the echoes ring with their glad shouts. There, on board, was the food to fill the mouths of wives and little ones, and warm shoes and clothing to cover their shivering bodies. What is shining

92

gold and silver worth, where there is nothing to be had for either? All now was peace and harmony, and plenty covered every board. Nothing unites a community like the sharing of each with the other in joy or sorrow.

'January 1st, 1851 opened mild and pleasant, but not anticipating New Year's calls I had not spread my table with tempting luxuries, but had sat down to think over the happy days spent far away, and of the many friends who perhaps would miss my hospitalities and greetings, when the door of my parlor-dining room- and kitchen (all in one) opened, and there before me was a group of laughing Indians of all ages, from the brave old chief May-je-ki-jik and his squaw, to all the little niches, and all the members of the tribe he could muster. As the outer door opened, all the other members of my family fled through an inner door and looked through a crack to see how I would receive my callers, but I had no time to arrange a program, for the old chief rushed up and greeted me with a kiss, and all the rest followed his example. One young brave[88] had painted his face to indicate he was in love instead of having an engagement ring to proclaim the fact. I cannot tell exactly how the red paint was put on, but it was in lines pointing to his heart. His long black hair was braided and hung down the sides of his face, and braided in it were small brass thimbles strung on a soiled pink ribbon, and when he moved his head they produced a tinkling sound. The old chief appeared in his accustomed blanket and embroidered leggings and moccasins, and his wife had on a rather scant broadcloth skirt, elaborately embroidered with porcupine quills and beads. Fortunately I had plenty of good substantial food to set before them, and they went away satisfied. Of course I felt honored, as I should that such distinguished guests had put on their best attire to call on me. Every New Year after that for a number of years I spread table for them, and they never failed to come and "eat salt" with me, and I have always had their friendship and good will. [This friendship continued even after the death of the "old chief" in 1862, being carried on by his daughter, Charlotte, (Plate 25) and her husband Charlie Kawbawgam. Plate 26 is a unique photo of the chief in everyday clothes, taken about 1896 by Everett's grandson, J. Everett Ball.]

'Later in the day Mr. Jed Emmons, of Detroit, and Mr. R. J. Graveraet and Captain S. Moody called. I had rather suspected that these two gentlemen had induced my first callers to pay me their respects but they disclaimed all knowledge of it. Mr. Graveraet informed that the Indians always made it a point at Mackinac and the Sault to call upon the white people, and probably the fashion had reached here.'

These 1879 memoirs added a tribute to Marji Gesick's friendship: "The brave old chief sleeps with his fathers in an unknown grave,[89] and of him much praise could be said, but I must leave him for some future time, to come down to the present, January 1st, 1879."

For the dramatic story that happened next, we go to Aunt Ella's letter: "Now I come to my little story: -- That first winter in our little cabin was a terrible one for our poor mother. She slipped on the icy steps at our front door and fell breaking her leg and the <u>next night</u> February 22nd Kittie was born."

C. R. Everett interjects "I was told by one of my aunts and think it was the same Aunt Ella, that grandmother had to do her house work, washing and etc. before they took her to Detroit, with her leg resting on a chair."

Aunt Ella continues:

We had an old country doctor there but he did not do his job very well, so, when the first boat came in the spring, Mother with Emma and baby were sent to Detroit for treatment. That poor leg had to be broken and set properly. Mother did not get well enough even for crutches until navigation closed so she had to stay there all winter.

Emma was put into a boarding school [in Detroit or Jackson] and Charlie & I lived with Aunt Mary [Johnson]. Father took Edward up to the mines with him where there was a boarding-house camp. I was only 6 years old and Charlie two years older. We were great pals and I wanted to do everything he did. Of course we had horses to drive to the mines, but many times Father walked on snow shoes to the mines in winter and only came home once a week.

Water had to be brought from the Lake in barrels for household use, and many times per day we children could run to the Lake and bring small pails of water for drinking, and what fun that was. All summer we lived on the beach. So that little cabin was our first home in Marquette.

The Everett Recollections continue[90]:

That building was removed to give place for the Burt block (Plate 20) and placed on the west side of the manufacturing shop, as a paint shop, on Main street [just west of Hagar & Johnson's factory], about half way from Front street to the first depot of Marquette. [When torn down on January 31, 1888, it was written up in The Daily Mining Journal of February 1 and February 6 as the oldest house in Marquette. Its hewn timbers were "found as sound and as firm as the day the house was built".] That building escaped the great fire of 1868, which took every building I had in Marquette, and every building on Front street from Superior to Michigan street. [See later description of the great fire.]

## The Second House

Clearly that little four-room cabin of hewn timbers had barely sufficed for the first winter. Not only did Everett have this large family to house, but, as the leading citizen, he needed to provide for civic activities and for visitors from "down below". So, besides working at the mines during the year he was batching, Everett and Johnson were busy. The sawmill that they had built on their earlier trip enabled them to start on the first steps of a continuing building program. As Everett reminisced: "I built my dwelling that Winter. When my wife returned in the Spring ..., my house was ready to receive her."

Aunt Ella's letter (quoted above in part) describes it: " ... quite soon Father built on that same ground [that she had described] a very large house & a brick Bank on the southwest corner [of Front and Main Streets] ... .That new house had to be very large as Father often had to entertain the rich men from Boston [Cleveland ?] who owned the mines. Our big house with barn, ice house, chicken coop & yard, vegetable garden & flower garden filled that whole block."

The relief and pleasure that they all felt in leaving the little house was palpable. The new house was not only big enough for the family, but it allowed the Everetts to exercise their talents for hospitality, a gift that was much needed and much appreciated in the new settlement. In that raw and unsettled land, housing was a problem from the first. Kawbawgam's log "hotel" was obviously inadequate. It gave way to one run by A. N. Barney, who had left the Forge and who would eventualy succeed Everett in a long career as Probate Judge. But when an appropriate welcome to an important visitor , like "the rich men

96

from Boston", was called for, the Everetts happily stepped forward.

Their hospitality was only a part of their works and it was certainly broadly given and not limited to the rich and famous. As Edward Fraser wrote some 28 years after the events, although he was a stranger, "you opened your house, not only for me but to other young men". After elaborating at length on their kindness and "social and moral influence" he still had to say "This and much more can be said of you".[91]

Without restricting their businesses, Everett and Charles Johnson turned to many civic activities, establishing local government. Even while they were still settling their affairs in Jackson, Johnson was named Township Road Commissioner.

Although Marquette County had been established by the legislature in the mid-40s, no elections were held until 1851. At the first election on November 4, 1851, by the unanimous action of all 62 voters, Everett was elected Probate Judge and held that position through 1856. Charles Johnson was elected County Treasurer from 1851 through 1856. Everett was also a Supervisor of the Township from 1851 to 1854 and presided as Chairman of the Board at the first special meeting of the board on September 13, 1852 at his new home. He also served on the first Grand Jury in the county on August 5, 1852, meeting in the office of Heman B. Ely.

None of these positions occupied all of his time. He was building houses and stores in conjunction with Johnson, who became known as "Uncle Charlie" and the principal builder of Marquette's first homes. Everett also had time for two major projects, described in later sections.

## St. Paul's Episcopal Church

It was in 1856 that the Everetts became the principal organizers of the Episcopal Church in Marquette. Everett was the first signatory of its Articles of Incorporation, dated August 18, 1856. At the organization meeting, he and Henry H. Stafford were elected Wardens and thereafter, as Senior Warden, Everett presided at the Vestry meetings.

At the meeting of November 3, the expenditures for the church building were reported, as well as the amounts collected to pay them. The latter were $1098.56, of which Everett donated $841.27.

At the same meeting it was:

Resolved, Whereas we have been informed that our worthy friend Mrs. Philo M. Everett has expressed a willingness to undertake the task of furnishing the Church with carpets, cushions for the reading desk, pulpit, &c. in case she has our approbation;

Therefore be it unanimously resolved that she has our unqualified approval as well as our earnest wish that her labors may be crowned with success;

Resolved that for her efforts in this matter she will be entitled to & shall receive our warmest thanks;

Resolved that the Secretary be requested to communicate these resolutions to Mrs. Everett attested by his signature.

The Secretary, Peter White, did so communicate.

Everett served as Senior Warden for many years, eventually relinquishing the position when he could no longer serve, to be succeeded as Senior Warden by his son-in-law, Dan H. Ball. Everett then became Warden Emeritus.


## The Mines in the Early Fifties

The Everett Recollections continue[92]:

Mr. Jones' management of the Jackson mine and forge was not very profitable. He got more on his hands than he could carry. [This was a charitable understatement. According to Peter White, the Company suspended operations in 1850, the Agent left in the fall and the men "talked seriously of hanging and quartering Mr. Jones, who soon after left the country."]

S. H. Kimball and General Curtis [of Sharon, Pa.], with other New York friends, bought up most of the stock [so that it was known for a brief period as the Sharon mine]; they leased the forge and mine to two brothers, Benjamin and Weston Eaton[93], of Pennsylvania, and they came on and took possession, bringing with them teams and supplies; but they found that four horse teams in this snowy country would not work. The snow path would not hold up more than what one team could haul. They made iron and carried on the works two or three years and then failed, losing all they had. [Fortunately, as noted above, Everett had disposed of his Jackson interest and was not one of those who 'lost a fortune in the forge operations', as a later history stated[94].]

The Graveraet and Harlow company went on making bloom iron for two or three years, but it did not succeed as they expected. Fisher became tired of furnishing money and nearly failed himself, having furnished over sixty thousand dollars. They finally failed and turned over their entire property to the Cleveland company.

It was now evident to both the Jackson and Cleveland companies that there was no money in making iron in an old fashioned forge, and they turned their attention to shipping the ore.

## Plank Road and Railroad

There now ensued a lengthy and acrimonious controversy about how to transport the ore from the mines to Marquette. A plank road corporation was organized early in 1850 and in the next year, John Burt made the first survey through the "36-acre plat". In 1851 the Jackson and Cleveland interests contracted to ship their ore over such a road and, after a new survey, work commenced. The plank road consisted of rough planks, laid side by side. Before it was put in use, light strap rails were sent up from Sharon, Pa., and it finally went into operation on November 1, 1855. Mules pulled cars carrying about four tons each and, when loaded, the bottoms scraped on the wheels and had to be trimmed.

In the meantime, Heman B. Ely of a Rochester, New York family became interested in the project. After an initial interest in the plank road, he turned to a railroad, which ultimately replaced the plank road.

Everett's Recollections[95] detail the story, including his participation in it:

In the spring of 1852, I went to the Sault, and met Heman B. Ely, with his engineers, coming to Marquette, for the purpose of surveying a railroad from Marquette to the mines, having made a contract with the Jackson and Cleveland mining companies to transport their iron ore from the mines to the lake at a much less figure than the ore has ever been carried since. His contract did not specify when the road should be finished, but it was to be pushed as fast as he could do it. The survey went on, and he built a dwelling, which is now a part of the Ely house, near Whetstone brook, and an office near by. The work progressed slowly but surely; after a time, the companies became impatient at his slow progress, but his reply was; 'I am doing as fast as I can pay up.' (He always paid promptly). But the Sault canal was to be built, and the companies were anxious to be in readiness to ship ore. Finally the Jackson company brought on some supplies for commencing a plank road, and the Cleveland company joined them. Ely didn't frighten, however, but kept steadily digging away. The parties had several meetings in my parlor. Ely said to them that he was doing as fast as he had means, but if they would furnish him the money he would push it on as fast as possible. They said to him, they would take the controlling stock and furnish the money. This he refused, but said he would let them have just one half and no more. That they would not do, and the meeting broke up. The plank road was begun and pushed with vigor, but the two companies soon collided, the plank road getting on to Ely's line, that was on record, and there were several suits between them in the District court. The ore companies had Mr. Walker, of Detroit, but Ely was his own lawyer, and Ely always came out ahead.

The plank road took possession of the wagon road whenever it was convenient, and in some places where teams could not get around the place they were obliged to travel on the plank road. One Saturday night Mr. Ely sent to the stable for his teamster, Plumtree, a burly Frenchman; when he came Mr. Ely said to him, he wanted him to take a load of supplies up to his men at the Eagle Mills in the morning. Plumtree said Himrod had put up several gates on the plank road at points where a team could not get around them and had put locks on them, and they were all locked up that afternoon. Mr. Ely was a rather slow and deliberate speaker; he said, 'Mr. Plumtree, I want you to take a load of supplies tomorrow morning to my men at the mills, never mind the gates.' This was spoken in such an emphatic tone Plumtree seemed to understand it and said, 'Yes, yes; I will do it.' Sunday morning Plumtree loaded up his team, taking his ax along with him. That was no unusual thing, for most of the teamsters in that day carried an ax to use in case of accident. He took the load up and returned. Monday morning there was quite a commotion in the plank road camp. Some one had cut down all the gates and thrown them off the road. Parties were at once sent out to find out who dared to cut down Himrod's gates. They brought word that no one but Plumtree with Ely's team could be heard of. He had been seen going up and back, but no one else had been seen on the road, and there was no doubt that he was the man. Mr. Plumtree was promptly arrested upon a warrant for cutting down the gates, and brought before the court, Mr. Ely appearing for Mr. Plumtree. Several witnesses were produced and proved beyond a doubt that Plumtree went up the road and back with Ely's team, but no one saw him cut down the gates, and the court discharged the prisoner. The gates were never put up again.

The plank road was finished and the strap rail laid in 1855. I took the contract for excavating what is known as the Jackson cut, in the winter of 1855-6 [actually 1854-5]. The Sault canal was to be opened in the spring, and the owners of the plank road were anxious to be ready to ship ore. I was obliged to work night and day to have it ready at the opening of navigation. The bed of the cut was twenty feet wide, with sloping banks. All was completed in due time, and the track laid through it onto their dock where the Grace furnace now stands. [Everett was said to have had charge of building the Jackson breakwater and dock and these must have used much of the rock from the Jackson cut.] The canal at the Sault was opened early in the summer [June 18, 1855] and forty or fifty mule teams landed at Marquette. They were put to hauling ore from the mines, but it was soon found the road was a failure. Generally the teams did not reach Marquette from the mines until 12 o'clock at night. Some one of them would break down, and that would hinder all behind it. They had no turn-outs, so as to pass each other and it was difficult to pass on one track, as has been found by all railroads. They divided their teams, stationing one half at the mines, meeting at Eagle Mills; but the road was a failure. More ore could have been hauled on wagons without the rail.

## Everett's Commercial Ventures

While all this was going on throughout the early fifties, as Everett says in his memoir, "I employed myself at various occupations for several years. I built three stores and an Office. ..." He described himself throughout the fifties as a "Commission Merchant", acting as he had at Jackson as a traveling agent to import into the pioneer community its necessary supplies. Local directories list him, separately or with others, in various enterprises. A merchandising partnership with Sidney Adams (a relative of the Huntoons) continued until 1858 and they were advertising lime, coal, blasting powder and hair for plastering. But Everett was not as interested in retail merchandising, as he was in larger transactions.

**PLATE 21**
**Sketch Map of the Great Lakes**
*(Portion of map, Plate 17)*

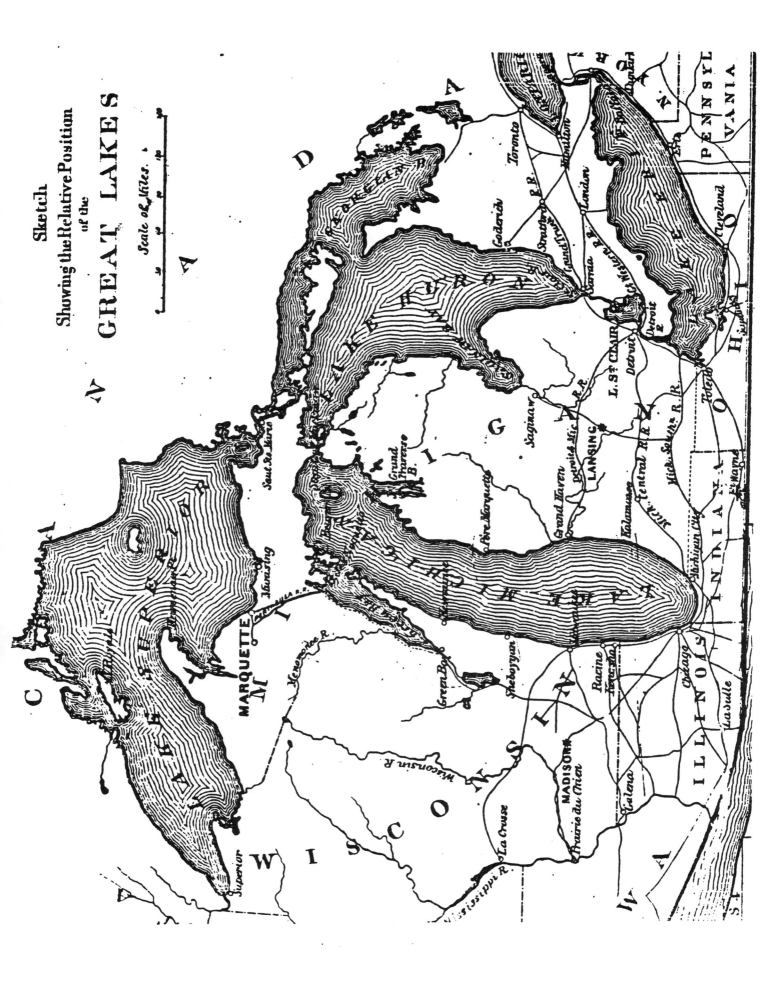

Sketch
Showing the Relative Position
of the
GREAT LAKES

Scale of Miles.

He traveled considerably. If it hadn't been apparent before, it should be re-emphasized that travel to Marquette from outside was not difficult -- it was virtually impossible for about half of the year. From December until May or June, navigation closed on Lake Superior. Yet Everett was at the Sault in the spring of 1852 and in Cleveland and New York in the winters of 1857 and 1865. Not only in winter and spring was travel hazardous, even in summer and autumn, there were several instances of boats lost without a trace. When there was snow, only the occasional mail could travel on dog sleds driven by the Indians to Green Bay, where there was finally a rail terminal.

In February, 1855, Mrs. Everett wrote[96] to a friend to hail the new link to Lake Michigan (where boats could come earlier and where there would soon be a railhead, as shown in Plate 21):

The long talked-of road from here to Bay de Noquette is now opened, and the first team arrived here yesterday with sixty bushels of grain, the first load of anything ever brought by the overland route to Lake Superior. Hitherto dog-trains have been employed to transport the mails; and one enterprising citizen succeeded last winter in transporting a few barrels of pork by the same means.

Those were the days of small things, not to be despised, for without the inventions of the untutored Indian, we should now be scolding about the postmaster general not sending our mail, which we had very good reason to do in the early part of the winter.

There have been up to this time over sixty bushels of mail matter drawn by dogs brought to this place; for the future there will be sleighs for the convenience of travellers and the mail *once in ten days.*

PLATE 22
Map of St. Mary's....Iron Lands,
ca. 1857
*(Note that it shows the Plank Road and the Railroad going to
a single dock, also the Railroad going beyond the three mines.)*

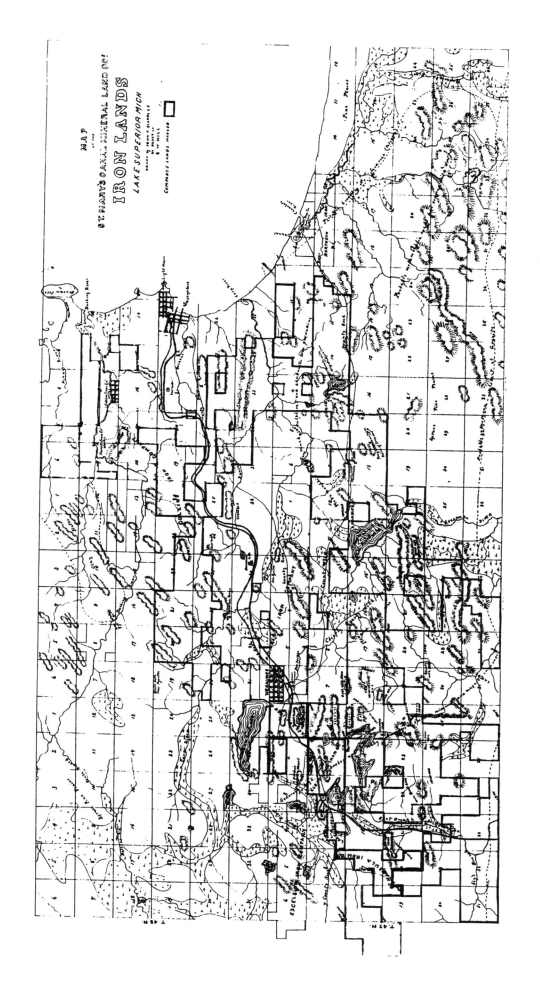

MAP
of the
ST. MARY'S CANAL MINERAL LAND CO?

IRON LANDS
LAKE SUPERIOR MICH

You will perceive that to Marquette belongs the praise of having persevered and overcome every obstacle in her way, and the honor (if there is any) of having outdone every other town on the lake, being the only one in communication with the little world outside this desirable region.

Detroit people had better be looking out, or Chicago will usurp the trade of this whole upper country. With a railroad from this place to Lake Michigan we can laugh at Deroit; and hie for Boston or New York without so much as saying, 'By your leave.'

## The Railroad Comes to Marquette

Now Everett tells[97] of the success of the railroad from Marquette to the mines:

In 1856 a grant of land was made by Congress to aid in the construction of railroads from Marquette and Ontonagon to the state line, which included the road from Marquette to the mines.

Senator Sumner was at my house the year before, and he said to me that whenever Lake Superior needed anything from Congress, if I would write him, he would do all he could for us. When this land grant was asked for, I wrote him, asking him to do what he could for it, which he cheerfully did.

Heman B. Ely died in the fall of 1856, but the railroad was pushed through, and completed to the mines in the following summer.

PLATE 23
Detail of Plate 22

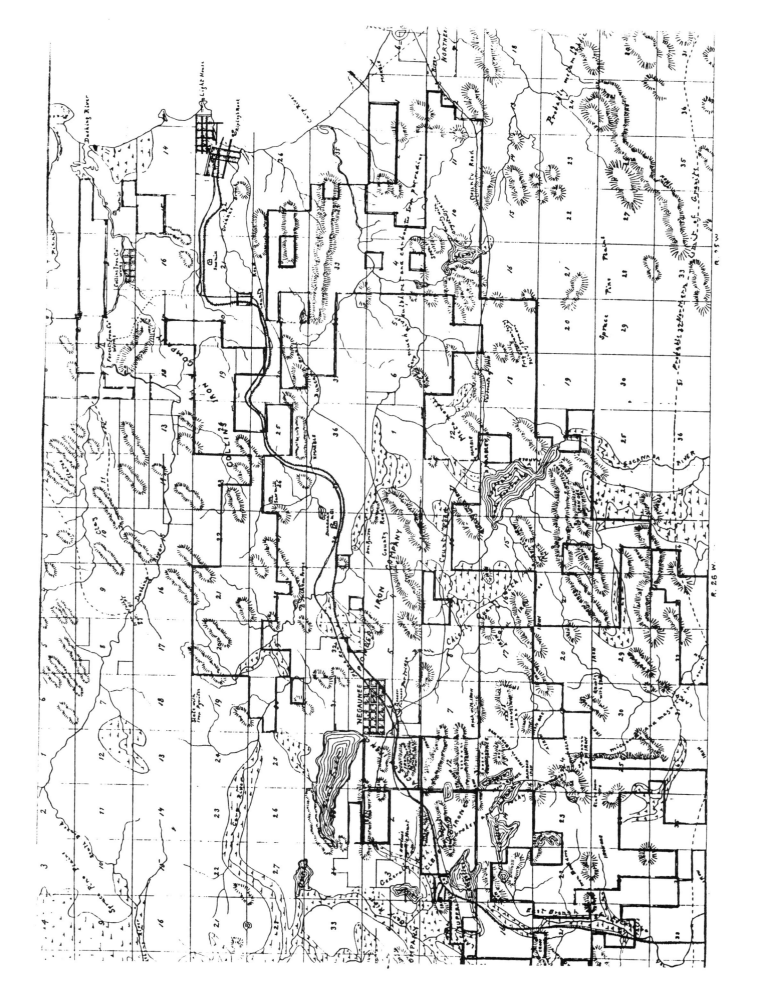

In the winter of 1857 I was in Cleveland, and the Ely's telegraphed me to charter a vessel to take a locomotive from Buffalo to Marquette in the spring. I was greatly surprised to find that there were only two vessels large enough, the E. C. Roberts and the De Soto. I chartered the E. C. Roberts, which was then in Buffalo.

I left Detroit for Marquette the first day of May, 1857, on the steamer North Star, and did not arrive at Marquette until the third of June. The steamer had to force her way through ice for six or eight miles. The last piece of ice was seen floating in the bay on the morning of the third day of July, 1857.

Everett summarized it: "In this summer [1857] Ely landed the 'Sabastapol', the first locomotive ever landed on this shore."

He continued:

A company of men to commence the Pioneer Furnace came on that steamer. That summer Dr. Ely came to take the place of his brother Heman in building and running the railroad. When the locomotive reached Eagle Mills, the ore companies made arrangements with Dr. Ely to take their ore at the mills, while they delivered it that far with mules. After the locomotive reached the mines the ore was all shipped over the railroad. The mules were sent away, the strap rail taken up, and the plank road was abandoned to the public and made a very fine wagon road. (See Plates 22, 23 and 24.) From this time the ore companies made money rapidly, mining and shipping ore. The 'Little Location', as it was first called (now known as the Lake Superior Mine), in 1886, just thirty years from the start, produced over two hundred thousand tons of ore. The whole product of the Lake Superior iron mines, in 1886, was three million, four hundred and sixty thousand tons.

PLATE 24

Dumping ore into rail cars at the Jackson Mine, ca. 1857

## The Kobogun (sic) Litigation

A controversy, dealing the share given to Marji Gesick[98] in 1846, arose after the Chief died in 1862. (The Supreme Court was, however, vague about the date, pointing out that he died "at a time not ascertainable by reason of the inability of the Indians to fix it definitely by our methods" 43 N.W. 603.) After he died his daughter, Charlotte, the wife of Charles Kawbawgam, the next chief of the Chippewas, found the paper that gave her father a share in the mine and took it to Mr. Everett and asked him to see The Jackson Iron Mining Company with regard to obtaining something. (See Plate 25.) The Court described this step:

> This paper was retained by Marji Gesick, and was found some time after his death in a box containing some of his possessions by his daughter Charlotte Kobogun. Neither he nor any of his family or relatives could read or write. Francis Nolin [Louis' son, aka. Jacques LePique], a Cree half-breed, who had married a Chippewa woman related to Charlotte, testifies that they were talking about the iron location, and Kobogun showed this paper, which they thought related to it, and, for the purpose of finding out, showed it to Mr. Everett, one of the original parties who had dealt with Marji Gesick, and Mr. Everett made an endeavor to get matters righted in her behalf.

By this time Everett's business was taking him to the East, where he and his family were planning a trip to celebrate Abraham Lincoln's Second Inaugural, now that the great Civil War was finally coming to a conclusion. It was on this trip that Everett took the paper to New York and called on Mr. Stewart, the president of the reorganized Jackson Mining Company in order to settle it. He described how he took the paper and why he remembered the date in the testimony at the trial as follows:

**PLATE 25**

Charlotte Kawbawgam and daughter, ca. 1870's.

*Charlotte Kawbawgam was the daughter of Chief Marji-Gesick, and wife of Charlie Kawbawgam, whom she married in 1851. It was she who pressed for the share in Jackson Mining Company given to her father.*

Q.  Did Charles Kobogun and his wife ever bring this paper of May 30, 1846, to you and request you to see the Jackson Iron Company with regard to obtaining something for the paper -- obtaining their rights under it?  A.  Yes sir, I suppose that is the paper.

Q.  You know it to be the paper, don't you, in looking at it?  A.  I should think it was the paper. [He had indicated elsewhere that his eyesight was failing.]

Q.  Did you take that paper?  A.  Yes, sir.

Q.  What did you do with it?  A.  I took it with me to New York, and called on Mr. Stewart, the president of the company to settle it.

Q.  What year was that?  A.  1865.

Q.  Have you any way of fixing the date in your mind -- any other circumstances that happened by which you can be positive as to the date, or the year?  A.  Yes, sir. I spent the winter in New York mostly -- New York and Washington.  It was just before Lincoln was killed.  I was in Washington a few days before he was killed, saw the president and had some business with him [See further discussion below.], and that was while I was staying in New York that winter.  I called on Mr. Stewart as soon as I went to New York, I think it was in January, 1865.  I wrote to Mr. Stewart before I left here.

Q.  And you say the spring following Abraham Lincoln was assassinated?  A.  Yes, sir, it was the same winter.

PLATE 26
Charlie Kawbawgam, 1896
(*A unique photograph by J. Everett Bull*)

Q.   Shortly after you saw Mr. Stewart in the spring?   A.   Yes, sir, it was in the winter before Lincoln was assassinated; I think it was in January that I saw him. You asked me for some incident that would refresh my recollection.   Going to Washington and seeing the president just before he was assassinated was one item.   That is what makes me think that I could not be mistaken in the year.

Q.   Well, did he make any offers of settlement? A.   He did after a while.   He did not immediately.   But he asked me to come down to their office in Trinity building and look the books over.

Q.   You say that he asked you to come down and look over the books.   What books was that?   A.   The Jackson old books that we had first.   They passed into their hands, and they took the books to New York.

Q.   He asked you to come down and look them over? A.   Yes, sir.

Q.   Well, what then?   A.   He wanted to see what the books said in relation to accepting this bargain that Berry and Kirtland had made.   I told him I thought it was on the books, and we went down and looked the books over and found that the company acknowledged the claim, sanctioned it, and appointed a committee of ways and means to ascertain how it should be settled, but we could find no report of that committee.

Q. You say he afterwards made an offer. What did he do?  A.  Well, I rather urged him to settle it up; the parties were poor and needed it, and they ought to have it; that they were entitled to something, and I thought he ought to do something, and he said if they were poor he would give them a hundred dollars, but he didn't believe the claim was good for anything.  I asked him why, and he said because the company was not the same company; they had changed; it was not the same members of the company.  I told him I would not take a hundred dollars; if it was worth anything it was worth more than that.  I wouldn't take it.  I had no interest in it, directly or indirectly, but I didn't think it would be doing the parties justice to accept a hundred dollars.

Charlotte Kawbawgam thereafter obtained more help and sued for her rights. The record, including Mr. Everett's testimony supporting her, is a priceless and nearly contemporary record of the circumstances.[99] The case went to the Supreme Court three times and created a split between its most famous justices, Justices Campbell and Cooley. The latter took the strong view that the daughter of a polygamous marriage, or perhaps no marriage at all, could not be recognized because she did not meet "the test of the laws which govern Christian nations.  But this test my brother Campbell does not accept." Justice Campbell cited a good deal of common law as to the validity of marriages made in accordance with local law but, more important, stated that the Indians were recognized by the United States as tribes capable of regulating their own internal affairs.[100] So in the third appeal, Charlotte joined with two other descendants of Marji Gesick, as urged by Judge Cooley, and they were recognized in their claim for the capital and the dividends with interest on the back payments that went back to 1861.

## Everett's Businesses in the 1860s

Everett had already built several buildings, including the first large office building, Everett's Block. Now he cooperated in the building of an even larger building. In 1858, the little first house was moved so that the large brick and masonry Burt Block could be built on Front Street. The little house became a warehouse and was the oldest house in Marquette until it was finally pulled down, as described above.

Of the next phase, Everett says he "finally drifted into Banking, selling village lots, collecting rents, paying taxes and insurance. I had a large Family to supply with food and rayment. We had no delivering goods in them days. I had to select and carry home what was wanted." These were natural developments, considering the thirty-six acre plot that he and Johnson had entered (even though they had shared it with Cleveland Cliffs) and the substantial building that they had done. For example, in 1864 the Watsons reported that "Uncle Charlie" was building a house for himself, next door to that of Mr. and Mrs. Ambrose Campbell.

By the 1860's Everett was listed as a banker and insurance agent. Competition grew up. He represented Aetna Insurance Company, but Peter White's biographer[101] tells how White, having left a law and insurance firm with Matthew H. Maynard, concentrated on an insurance agency after 1865, which he apparently wanted to monopolize the entire insurance business:

Although the agency began by representing but two companies, by 1862 it had five. Shortly after White took the agency, it represnted ten companies; but he was not yet satisfied. Philo M. Everett represented the Aetna Insurance Company, an agency White wanted. The Aetna was among the first large insurance firms extending fire insurance to the

Midwest. It had a good reputation and had surpassed other companies in assuming risks in relatively unknown areas like Marquette. After Everett refused to sell, White wrote the company and obtained the agency in August, 1866. Thereafter, he paid Everett $1,000 for his loss, even though he was not obligated to do so.

Prior to this White had exhibited a great deal of animosity toward Everett[102] but, after he had what he wanted and could condescend to make a "gift", relations seemed to have improved. Such was the social tolerance of Marquette, where the early miners were not above claim-jumping (as we have seen), but then went on to become respected philanthropists in the city, that the two families went on as social friends. (Another historian, however, tells of a family that did not take one of White's tactics with such equanimity. According to a legend that he relates, they reached back into their Indian heritage and pronounced a horrible, and seemingly effective, curse on White and his family.[103])

## Everett's Home and Family.

Everett began banking in his office building. His son Charles (born in 1843) became a cashier at a very early age and was eventually listed as a partner. The business prospered and his home grew. Among the furnishings they acquired, Mrs. Everett was reported to have highly prized an original portrait of Sir Walter Scott, the famous British author. It was described as a very fine portrait, painted when Scott was in the prime of life. Mrs. Everett also set a beautiful table with imported china and coin silver made for them by T. Leavenworth, a leading silversmith of Detroit.

A new person soon joined the family in the person of Dan Harvey Ball. He had studied law at the University of Michigan, but came to Marquette in 1861 to take care of a mercantile business that his deceased brother had started only the year before. He wound that up and, with Alexander Campbell, became publisher of the Lake Superior News and the Lake Superior Journal, which later became the well-known Marquette Mining Journal. He soon became Registrar of the United States Land Office at Marquette and held that appointment until after the death of President Lincoln. It was in 1863 that he married Everett's eldest child, Emma. He resumed the practice of law in 1865, at first in partnership with Marquette's first lawyer, Matthew H. Maynard. Then, for a few years the young couple moved to Houghton, the center of the booming Copper Country. After four years they moved back to Marquette, leaving behind a partnership in charge of his cousin, Joseph Harvey Chandler. Thereafter D. H. Ball practiced in Marquette until his death in 1918.

Although travel was so difficult, especially in winter, there were more trips besides the trip to the Sault in 1852 and wintering in Cleveland until June 1857. Emma Everett (later Ball) tells how they traveled once, apparently from a trip to Chicago to buy her trousseau, back to Marquette in 1862 on a construction train, as the railroad was just reaching Green Bay. From there they took a boat to Bay de Noquette (aka. Bay de Noc) and then a stagecoach to Marquette.

By 1864 the Peninsula Railroad was completed to Lake Michigan and that was the improved route that Everett took to spend the winter of 1865 in New York and Washington. His wife and the Balls were with him to meet President Lincoln and attend the Second Inaugural. It was then that the President reappointed Dan H. Ball as the Federal Land Agent.

The summer communications on the Great Lakes had been sufficient to bring Godey's Ladies Book to Marquette and so the ladies had almost the latest Paris fashion for the Inaugural Ball. To the disappointment of Emma Ball, she became too ill to attend the Ball and thereafter could only point to the fragment of her gown that was incorporated into a silk-covered quilt.

## The Crisis of the Great Fire

Everett's business was large in 1867 -- he said that it grossed over $700,000 and netted $2,500. (The net does not seem large to us but it would be, if converted into today's dollars.[104]) Everett said [105], "My pen did the whole of it, for I had no clerk. I had also a horse to care for . I worked early and late. In the summer I went to my Office as soon as it was light and posted up my books before breakfast and working to nine and often to twelve at night." He said, "I was strong and healthy, my wife working equally hard, but we were making money and happy, never once thinking that I was using myself up." {Like his report of the arduous trip to Lake Superior in 1845, this memoir refutes the tale of early ill health that finally led to his blindness.)

In the Spring of 1868, however, a sudden breakdown did occur. As Everett described it: "One night at twelve oclook I went to my bed feeling strangely bad. There was no sleep for me that night. In the morning my wife found me delerious. She rallied the doctors at once and in the course of the day my reason returned but my hard work was forever done". He engaged a clerk, decided that he and his wife must have rest and they left on a vacation of unstated duration.

They got only so far as Kalamazoo in Lower Michigan, where his brother Cyrus lived, when another disaster struck. Everett's memoir says:

A telegram reached us saying that Marquette was in ashes! Our all above the ground was gone, except a horse -- Stores, Banking Office, Dwelling and Barn with all their contance gone -- Even the clothes on the Children's backs was destroyed, trying to save something. That night when the Family went to bed there was nine beds made up, mostly new, my Carriage, two pleasure sleighs, oats, hay and all the parafanailia belonging to such an establishment gone to ashes.

The headlines and reports in the papers[106] emphasize the general havoc: **"Marquette in Ashes ! —Terrible Destruction of Property ! —An Entire Street Destroyed ! ... Only 2 Stores Left Standing !"**.The paper detailed how every place of business on Front Street was destroyed and named them -- from Superior Street north to Washington Street, more than three blocks, on both sides of the street, consuming Everett's Block and the even larger Burt Block and running down to the Lake, burning the three large docks.(See Plate 27.) Heroic efforts finally stopped the fire and saved some of the contents of the burning buildings.  Bank valuables were saved by putting them on a small boat in the harbor.  Frantic efforts were made to save the contents of the drugstore, especially because there and on the docks there were so many inflammable and explosive stores. In Everett's Block only the Township Records were saved, and the Township Library as well as the library and papers of the Judge of Probate were lost.

**PLATE 27**
Detail of Plate 17,
marked to show area burned.

The short telegram caused an immediate reaction in far-off Kalamazoo. Everett described it:

That was a dark night to my Wife and my self We went to bed but not to sleep. It was the hardest night we ever spent. When we ware begining to think of rest in our old age, broken down with hard work, trying to accumulate something to support us and have something for our Children to begin the World with and not have to commence the World with out a dollar as I did.

The first question we asked each other after getting into bed was, 'What shall we do?' It was a hard question to answer but it must be answered -- I had kept my own books and I knew how I stood. My real estate was clear from any incumberence; I was not in debt; I could settle up and have four thousand dollars left, two blocks of land now clear from buildings, one 160 ft. front, the other 110 feet front with a lot of vacant Village lots.

Our first thought was to come back and sell out, but on consideration we saw that every one would wish to sell and there would be no buyers.-- My taxes would be five or six hundred dollars. How was I to pay that? My health was such that I could not labor and, if I could, I could not pay that amount and support my Family. Then to clothe up my Children and procure material for keeping house would use up the most of my four thousand dollars. Then I would have no money to carry on my banking business.

**PLATE 28**
Everett's Block
ca. 1996

We worried over it all night. At last I made this suggestion -- to go East to Hartford and by the aid of my Friends obtain a loan. With that I could come home, go on with my banking business, build a block of stores, pay cash for every thing -- ask no credit of any one. That would give me a credit at home and abroad. By the time my money was out, my stores would be bringing in a handsome rent. That plan was settled on.

In the morning I started for Hartford and obtained fifteen thousand dollars. We came home and started the building at once. The whole plan worked as we had suggested -- but I did not buy furniture for house keeping. I wanted the money in my business. I boarded my Family eighteen months to save the advance.

All Marquette was rebounding with amazing vigor and speed. Because the fire had occurred early in the shipping season (June 11, 1868), materials and labor were brought in by water from "outside". Immediately after the fire, the Marquette Village Council prohibited any wooden building in the central, burned-out area. This wise requirement no doubt slowed rebuilding, but laid the foundation of sound reconstruction. Indeed, it prevented a recurrence when a fire broke out in 1891 and threatened the very Block that Everett had rebuilt. That building was so well built that it was modified several times and was awaiting a new reconstruction in 1995. (Plate 28)

Fortunately, local building materials were available. A quarry just south of the town had been operated for ten years by the Burts, who had used it to build a quarry office (which is still maintained as a "pioneer home"). Brick was also available. No doubt the mining interests up the line were quick to provide their resources of men and material for rebuilding the ore docks, the lifeline of the economy.

PLATE 29

Front Street, about 1872, showing Everett's Block re-built and his later house at the top of the hill.

Everett was back promptly to carry out his plan and seems to have succeeded with great speed, despite the difficulties and the ill health which he seems to have surmounted. He wrote, "I had a very serious time in building. It stormed most of the time in October. For two weeks the masons did but half a days work." But he carried through his plan and promptly restored his rents and the banking business.

As early as September a dear friend, Dr. McKenzies, could write from New York [107]: "The rapidity with which Marquette is being rebuilt shows the substantial nature of its business resources, as like the Phoenix, from its ashes it is resuscitated." He went on to congratulate Ella upon her approaching marriage and repeated his sympathy for their losses, saying "I have certainly felt much the loss of your painting of Sir Walter Scott, a relic not to be replaced in this country, and which in Edinburgh would have been worth its weight in gold. Often have I gazed upon it as with benign aspect, he looked down upon your family circle." He also found the death of Bruno, the family dog, "quite touching. Whenever any firing or extraordinary noise occurred, he always found a refuge under the dining-table, and there the poor brute met his fate." Before closing, Dr. McKenzies included a poem on "the dear old house ... Bruno ... and the relic dear of Walter Scott".

Aunt Ella wrote that for that period (through 1869) the family "boarded at Tibbetts House (where the Matthews house is now [in 1929]); from there I was married [Sep. 29 1868] and Kittie was sent to Emma Willard school in Troy, N.Y. [Earlier] I was sent there for one year".

Philo Everett Home Marquette
Corner of Front and Ridge St where
The Peter White Library now stands.

PLATE 30
The Everett House,
after 1870.

Although Everett's losses in the fire had been almost complete, he recovered rapidly. Probably in addition to the $4,000 that he mentioned, he had (according to the newspaper account of the fire) $6,000 in insurance on losses of about $20,000; this failure to carry adequate insurance was shared by the Burts, the largest losers, and others. In the census of 1870 Everett listed the value of his Real Estate as $79,500.

It was in 1870 that the family moved into their mansion, a beautiful Mansard-roofed house at the top of Front Street and at the corner of Ridge Street, where the Peter White Library now stands. (See Plates 29, 30 and 31.) The Dan Balls returned from Houghton and in 1871 built further along Ridge Street, at No. 411. (See Plate 32.) By avoiding the fire, they had a small share of the surviving Everett and Johnson heirlooms.

PLATE 31
Front Street with the Everett House.
(Courtesy of the Marquette County Historical Society)

## The Rise and Fall of the Banking Business

"Wildcat banking" was an abuse prevalent before the Civil War and in this period some steps had been taken to eliminate it. Nevertheless, private banks were still common and had an especially important place on Lake Superior.

Everett's prosperity at this time stemmed mostly from concentration on his private banking business and discounting "iron money" rather than from his rents. Peter White described the use of "iron money" in the mining areas to a Congressional Committee in 1875:

During a large part of the year navigation is closed and all contact with the outside world is cut off. The offices of these companies [copper, as well as iron, mines] which handled financial matters were located in the east — Boston, New York, Cleveland, etc. The product was sold in summer to eastern and south Midwestern cities. During the winter the payroll used to be met but often it was not possible due to the expense and distance at which currency was available.

Under the circumstances it became almost a universal practice to pay the laborers with a draft on the treasurer of the company .... Such drafts did not remain in circulation long but were collected in a store or bank and sent to company offices for collection. These drafts were not intended to circulate as money but their use by laborers and those who received them was unavoidable in the absence of other circulation.

The recent connection by rail and communication has already made the use of these drafts negligible.... 108

131

PLATE 32
D. H. Ball's Residence
1886

White was a large factor in the purchase, at a 10% discount and collection with 10% more for interest, of this "iron money", which was issued by many businesses as well as the mines. It was a large part of the volume of all the banks, including Everett's. Each bank required large capital to carry this business during the long closed season in winter and spring, when communication was shut off.

In this situation one of the Panics (that recurred so often before the advent of the Federal Reserve system) struck suddenly and viciously. An excellent history tells how it spread within 48 hours of the failure of Jay Cooke & Company in Philadelphia on September 18, 1873. Whether "in the backwoods of Michigan" or elsewhere "by the end of 1873 more than five thousand commercial enterprises had gone under".[109]

Thus it is not hard to see that, when the panic struck and mines were shut down, banks and their owners suffered heavily. White survived with (in White's own words)[110] "brass, impudence and conciliation", plus certain manipulation between his public and his private bank. But his losses before the depression ran its course were estimated at $690,000, and it took a Special Act of Congress to dispose of a tax claim that might have been even larger.[111]

Everett's bank, on the other hand, was completely ruined, but there was no breath of scandal or even criticism about it. We have found no report of the amount of Everett's loss and none of his depositors showed any. That was because he and his son-in-law, Dan Ball, quietly dug down in their own pockets so that none of their friends or depositors would lose a cent. It was probably an exaggeration when Aunt Ella said that he was left penniless. Although Eastern colleges for the daughters became out of the question, Everett still called himself a banker

**PLATE 33**
Detail of *Map of Marquette Michigan*
(August, 1888)
*(Note that it shows EVERETT'S BL on W. MAIN and his residence and
outbuildings at the corner of W. RIDGE and N. FRONT.)*

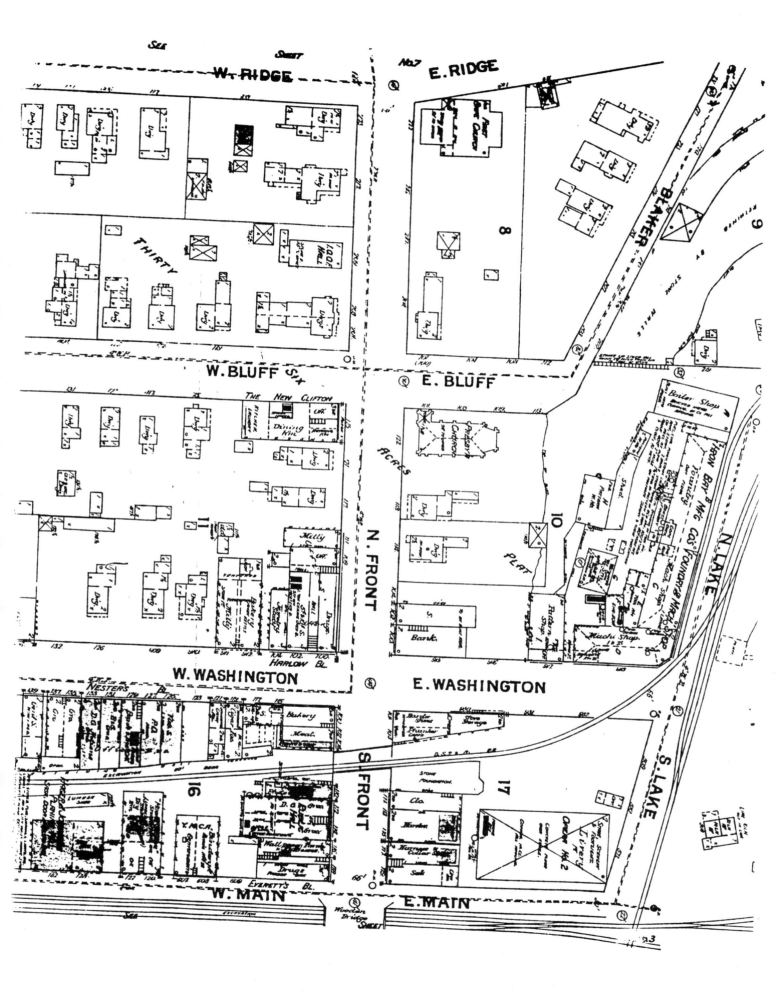

for a few more years and continued to live in the large mansion.
(See Plate 33.)

His son, Charles M. Everett, had gone into the bank
when he was "too young", according to Aunt Ella, and had become a
cashier.  Ultimately he was listed as a partner in "Everett &
Son, Bankers".  In 1876 he turned to the lakes and became Captain
of a tug boat, operating for a time to bring the large ore boats
into Marquette Harbor.  He had a favored position in this
operation because, from their mansion at the top of Front Street,
his mother could look over Lighthouse Point and out onto the
Lake. When she saw an ore boat approaching, she would hang a dish
towel in the upstairs window and her son's tug could be first out
to get the towing job.  (While Charles was still operating a tug
at Sault Ste. Marie, he became ill and returned to Marquette,
where he died on September 14, 1897.)

During the '70s Mehitable Everett (Plate 34) kept up
her correspondence and wrote several articles.  As of January 1
1879, she wrote, in the article previously quoted:

I have today looked in upon two groups of ladies in
silken attire, gliding to and fro over velvet carpets, with
eyes sparkling with pleasure as they meet and greet their
gentlemen friends and conduct them to their tables laden
with fruits from afar, and delicacies prepared by their own
hands; and I look upon these gentlemen bowing with grace
before those ladies, whose flashing diamonds add lustre to
the scene, (I beg their pardon) and I cannot help saying to
myself 'I wonder how many of them were crying out yesterday
against the hard times in Marquette and wishing they could
get away to a better place!'

**PLATE 34**
Mehitable Johnson Everett
ca. 1879

Well, that is well enough if they can find that better place; but energy and perseverance have made the town of today to contrast so strongly with the past, and the same energy can give it a future contrasting even more widely with the present. It devolves upon the young men of today to be up and doing instead of idly repining; and I wish the young of today could imagine or feel the privations others have passed through in their efforts to build up this beautiful city of ours. They would much more appreciate the Happy New Year which I cordially wish them all.

As Mrs. Everett's last quotation was intended only as a coda to the reminiscences about the 1850's, she did not begin to outline the many changes that had come to Marquette and all the Lake Superior country.

Marquette's lifeline was now the ore that flowed down the lakes in greater and greater quantities on giant ore boats through the ever-larger locks at the Soo.

As Mrs. Everett had predicted, the land routes straight south to Chicago replaced the lake route for almost everything else, particularly express and passengers. First, there were the sleighs she mentioned and stage coaches in the summer to Bay de Noc, soon to be replaced by rails. For nearly ten years there was a gap to the railroad from Chicago, which reached Green Bay in 1862. The gap was filled in by boat service in the summer and by sleighs after the ice was safe. Only after the first of several rail lines reached Marquette was the terrible isolation ended.

That changed first the business and banking world, and then the social life. The silks and velvets that Mrs. Everett described were part of the High Victorian decor. The simpler coin silver was replaced by elaborate designs from Tiffany and Gorham. With an overnight ride to Chicago on the Pullman cars, the ladies found all these things at Marshall Field's and the other stores.

The Congress Hotel, owned by the Kaufmans of Marquette, became a base of operations for all the visitors from the Upper Peninsula.

The "fruits from afar" were also provided by rail. For example, peaches, pears and grapes from the flat lands down under could now be had, as well as oranges and such from farther away. No longer did Marquette have to live through the winter on the apples, potatoes and root crops that could be put down in basements and outdoor root cellars.

Much as the gentle criticism of the new fashions may have been merited, it may perhaps have had a slightly bitter tinge based on the reversal of fortunes since the Panic of '73. Nevertheless, even as the pioneer life passed, the intrinsic Everett hospitality continued in the rooms and parlors of the Everett mansion.

To sum up the above-quoted article, Mrs. Everett's grandchildren wrote[112]:

> The mother no less than the father seemed to fit into the pioneer life, and was certainly a wonderful help and mate for her husband, and great assistance in many ways to the early settlers here. Even the original red man seems to have been attracted to her and her hospitality, ... With all the cares of a large family she never seemed to lack time to care for and entertain visitors, or to lend a helping hand in any way.

The Golden Wedding

On November 19, 1883, the Everetts celebrated their golden wedding anniversary, an event that was reported in the Marquette Mining Journal on November 24, 1883 as follows (with some deletions of poetry and other details):

Rarely is it permitted a wedded pair to spend half a century of existence in one another's companionship on this earth of ours, and the beautiful custom has obtained of celebrating such an event, when it occurs, by a 'golden wedding' on the fiftieth anniversary of the day when the nuptial vows were spoken by the pair in the bloom and freshness of their youth. Such a stage in the married life of our pioneer friends. Mr. and Mrs. Philo M. Everett, was reached on Monday last, and it was not allowed to pass 'unhonored and unsung' by their numerous family and society friends, about one hundred of whom assembled in the parlors of the Everett mansion that evening to shower golden gifts upon the venerable pair and assist them in suitably observing the fiftieth anniversary of their happy bridal day.

We have not space to enumerate the gifts in detail, but cannot allow a few of the more appropriate of these to pass without special mention. A purse containing $200 in gold coin, the gift of several Marquette gentlemen, was a fitting testimonial of the regard in which the recipients are held by the donors for their sterling qualities, tried and proved during a residence here covering the history of Marquette from its earliest settlement to the present day. The children and grand-children of the pair united in presenting them with a purse of gold coin, also, which contained $130 in gold pieces. In addition to these handsome gifts in coin, a large number of gold pieces of the higher denominations were presented by individual friends. The wedding cake, prepared by Mrs. L. C. Palmer [Emma Huntoon Palmer, a sister of Mary Huntoon Campbell], bore in its frosting a number of single gold dollars, and was greatly admired for its unique *richness*. The coins were the gift of Mr. Palmer. Another gift that had a tender significance which calls for more than passing notice was that of Georgie Ball, the grandson of the venerable pair. The little fellow, who is but ten years of age, had been furnished with money to buy such a present as might seem to him most proper. With a sense of the fitness of things beyond his years, he purchased an elegant wedding ring, and if he had any doubts of the wisdom of his choice before, they were dispelled by the delight with which his gift was received by his loving grandparents, and the approval of the assembled company.

Among the gifts was a small gold brick of native Superior gold from the Ropes mine which was the present of Mr. and Mrs. Julius Ropes, of Ishpeming. On one side is engraved "L. S. Gold," and on the other the name, "Everett." This gift is especially prized by the good couple, possessing, as it does, a historical as well as intrinsic value, it being from the first gold brick produced by regular mining for that metal in the Lake Superior region.

One of the pleasant incidents of this reunion of friends under their roof to celebrate the golden wedding of this estimable couple was the meeting of two school-mates of forty years ago, whose paths in life then diverged, not to cross again until this occasion brought them together. These were Mrs. John R. Martin, of Ripon, Wisconsin, a sister of Mr. Everett -- who was present with her husband -- and Mrs. J. [Joshua] Hodgkins [another sister of Mrs. Campbell], of Deer Lake. Forty years ago they attended the same school at Oneida, N.Y., and in all the intervening years they had not met until the event we are describing brought them once more together in the hospitable home of their life-long friends. There is much of joy in such meetings -- and, alas, a minor chord of sadness.

Marquette was but a site when Mr. and Mrs. Everett first settled here, to make their home where they have since seen a prosperous, populous city grow up around them. In 1845 Mr. Everett came here to seek for the iron deposit now known as Jackson mine, at Negaunee, of whose existence he had heard through Indian traders. He succeeded, after much diplomatic work among the Indians and traders, in finding the `iron mountain', and was chiefly instrumental in organizing the company which engaged in developing the mine. The history of his efforts in that connection are a part of that of the district, and as such so familiar to the great bulk of our readers as not to need recapitulation here. In 1850 he brought his family to Marquette, which has been their place of residence continuously since. When he came here, Mr. Everett was the only white resident of the place. As it gained in wealth, population and influence, he continued to be one of its foremost citizens in all measures looking to its social, moral and industrial advancement, until advancing years compelled his retirement from active affairs; and now, in their green, old age, he and his good wife are pleasantly traveling the downward road of life, sheltered in the affection of their children and grandchildren and the warm esteem of their friends. The rich usefulness and unobtrusive goodness of their lives deserves that their closing years should be full of peaceful content and happiness, and the MINING JOURNAL, adds, as its tribute to their worth, on this occasion, the earnest hope that they may be spared to participate in many future anniversaries of their happy wedding day.

This article is virtually an epitaph because Mrs. Everett died within a month of the anniversary, on December 11, 1883. The Mining Journal article mentioned `a minor chord of sadness", hinting delicately that Mrs. Everett was in precarious health. Nevertheless Everett felt her death as a sudden and shattering blow.

PLATE 35
411 East Ridge Street
Mr. Everett, his daughter Emma, two grandchildren, Mabel(?)
and George Ball, on front porch, 1890.
*(Courtesy of the Marquette County Historical Society)*

## The Last Years

He eventually sold his "big house" and moved into rooms in the Ball's home on East Ridge Street. (Plate 35) For a time he kept up some activity, investing with D. H. Ball in the laying out and developing of the town of Grand Marais, considerably to the east of Marquette, where the streets were appropriately named for members of their families. They were also interested in an old mine on Silver Mountain to the west, just over the line in Baraga County, but the miner they sent there in May, 1884, found nothing to follow up on.

By 1886 Everett's sight was failing. He was able to walk downtown but had too much sense of his infirmities to do it. His deafness kept him out of "common talk" but his mind was clear and alert. He kept up a correspondence with his dear "brother", Charles Johnson back in Jackson and reminisced with him about their first trips to the Lake. The vessels that they knew could draw only nine feet of water up the Soo River. Now the "lock at the Soo has thirteen feet and the new one ... is to be twenty feet -- it was a big load for a vessel at first to take six hundred tons of ore -- now they take from twenty to twenty eight hundred tons --then they got three to four dollars freight -- now they get one dollar and fifteen cents and make money at that." He went on to quote General Poe, the builder of the new lock, who said "these Lakes here do more than one third more business than all the Ocean business of the whole country.-- Few are ready to believe this but General Poe knows what he is talking about and he has the figures to prove it."

These reminiscences must have been gratifying. Like many others who had shared in the incredible growth of the area, he could look back with satisfied fulfillment. The iron mountain, which in the world of the 1840's was said "to supply the United States for all time to come', was long since consumed. Since 1840 there had been unanticipated demands for iron in thousands of miles of rails. for Civil War guns, for steel ships and, lately,

143

PLATE 36
Philo Marshall Everett
ca. 1890

for the steel girders of the nascent skyscraper age. As his 1887 Recollections and several biographers were to emphasize, Lake Superior tonnages of iron ore had grown from a few thousand to many millions of tons in about twenty years.

As a result of the mines, Lake Superior had transformed itself from frontier isolation into modern communities. On the Marquette lots that Everett had staked out, a series of buildings had been built, enlarged and rebuilt. Everett had to cajole men to come to a sub-Arctic wilderness, to bring in goods on little boats and later, when the Canal was built, to charter a Steamship to carry the first locomotive. Now many locomotives daily brought trains to Marquette and this rail service had changed the life of everyone.

Not only had trains brought the physical needs and the luxuries that Mrs. Everett had noted, they now supplied broad communications with a world that was no longer "outside". For example, the younger Everetts had been exceptions in going East to school. Not so with the next generation. Going to the colleges and the University in the Lower Peninsula and to others in the adjacent states and the East brought them into contact with the world from which their parents had come, but, of more significance, with the new developments at home and abroad. These generations would profit from new sciences, while engineering and professional schools would send back trained engineers, lawyers and doctors.

Ready communication brought not only immigrant laborers but new skills from abroad, exemplified by Swedish steel experts, British arts and sciences and other European industrial techniques. Becoming more cosmopolitan stimulated the community. Art, such as that taught by Everett's daughter Kitty, music and theatre were coming to Marquette.

Looking more inward yet, Everett could also view with satisfaction his increasing progeny of grandchildren, as well as relatives and friends whom he had brought to the Iron Range. And so, in summary, despite his infirmities, he could, like so many of his contemporaries who had also experienced these developments, look back and share the wonder expressed by Alexander Graham Bell: "What hath God wrought !"

It was early in 1887 that he finished and published the "Recollections" that are the real source for the early history of the Iron Industry. He must have had help in preparing the typescript that has come down in the family. Others may have guided him, but they seem to have made no substantial changes in the story, except for the omission of Marji Gesick, which had the result of giving to Louis Nolan some of the credit that belonged to the Old Chief. One biographer sought to explain this because of pending litigation. Perhaps someone urged undue caution, but that omission was certainly rectified when Everett, under oath, testified willingly and fully about the Chief's vital guidance.

Everett continued to write. The last memoirs, that stopped abruptly with events of 1868, are written on ruled paper perfectly legibly. Grace, the youngest of the Ball children, was at home until she went away to Wellesley in 1889. She read to him, reading over his memoir and his letters to him after he was so nearly blind that he could not read what he had written. She wrote, "I am more and more impressed with his patient acceptance of his blindness and deafness and the loneliness of being shut off from active life." She added in her letter to her cousin, C.R.(Ray) Everett: "He was certainly a grandfather to be proud of." (See Plate 36.)

## The End of a Long Life

On September 29, 1892, the Marquette Mining Journal reported:

### THE GRIM REAPER'S HARVEST

### Philo M. Everett Passed Quietly

### Away Last Evening at His

### Daughter's Residence

———

### FULL OF YEARS

The oldest resident of Marquette, the pioneer of Marquette county and of the Lake Superior iron country, the man to whom the Indians showed the great 'iron mountain', which became the Jackson mine, oldest of all the mines of the Lake Superior country, breathed his last at 9 o'clock last evening [September 28] at the residence of his son-in-law, Hon. D.H.Ball. So great a change is wrought within one short life time.

Philo Marshall Everett was born at Winchester, Connecticut, October 21, 1807. While a young man, he settled in New York State where he was married to Miss Mehitable E. Johnson, of Utica, in 1833. In 1840 he moved to Jackson, Mich., and engaged in mercantile business there, together with the forwarding and commission business.

Mr. Everett first came to Lake Superior in June, 1845, in charge of an exploring party sent out by a little body of men there organized into the Jackson Mining Co., afterwards the Jackson Iron Co. With this party he discovered and located the famous Jackson mine and after the summer here returned home. The next season he was back again to build the Jackson forge on the Carp river near Negaunee, the following year being spent at the same work.

From the spring of 1848 to the spring of 1849 Mr. Everett was in charge of the forge, moving to Marquette with his family in the fall of 1850. Here he had charge of the building of the old Jackson breakwater and in 1857 brought up for the Elys the first locomotive ever seen on the shore of Lake Superior. Afterward he engaged in the mercantile business here and later in banking and insurance, accumulating considerable property, which was swept away in the terrible depression throughout the region following the panic of 1871.

Mr. Everett was a zealous Episcopalian and was active in the establishment of St. Paul's church and in its maintenance and was senior warden until age and infirmity made it impossible for him to longer attend to the duties of the position. He also took great interest in politics, having been an ardent republican from the first formation of the party 'under the oaks'.

**PLATE 37**
Plaque in St. Paul's Cathedral

In the fall of 1883, shortly after the celebration of their golden wedding, Mrs. Everett died and since then in feeble health and with sight and hearing greatly impaired he has made his home with his daughter, Mrs. D. H. Ball. His other children are Mrs. B. P. Robbins and C. M. Everett of this city, Edward P. Everett of Grand Rapids, and Catherine E. Everett, now living in Chicago.

To the younger residents of Marquette and the Lake Superior country Mr. Everett was known in name only but in the earlier days of this city and county he was identified prominently with every movement to hasten the development of the iron range of which he was the father and to make known its resources outside and his name will always be an honored one in the Lake Superior iron country.

The funeral will take place from D. H. Ball's residence, No. 411 Ridge street on Thursday at 2 P.M.

A plaque (Plate 37) in St. Paul's Church in Marquette to Philo Marshall Everett and Mehitable Johnson Everett concludes: "[They] were the first communicants of this Church. He discovered the first Iron Mine in this county in 1845 and was Warden of this Church from 1856 to 1878 and was Warden Emeritus until his death."

# FOOTNOTES

---

1 Patricia L. Micklow, <u>Philo Marshall Everett and the Founding of Marquette, a Study in Historiography</u>, Northern State University (May 8, 1968), hereinafter "Micklow". Copy in Marquette County Historical Society (MCHS). Her study sets forth facts to show that Everett was the "actual founder of Marquette", as well as "responsible for initiating the major industry of the area, mining".

2 Justice John Voelker, pen name Robert Traver, <u>Laughing Whitefish</u>, (McGraw Hill 1965), also in paperback.

3 Some of the cases were titled "<u>Compo v. Jackson Iron Co.</u>"

4 This date (October 21, 1807) has been given in most family records, including several biographies, a church memorial, his obituary and the cemetary record. A consistent age was reported in the 1860 census. No primary birth data have been found.

5 See E. F. Everett, <u>Descendants of Richard Everett</u> (Boston, 1902). ( A copy is in the library of the New England Historic Genealogical Society and another is available to its members in its Circulating Catalog, which are referred to hereinafter as "NEHGS" and "NEHGS=C", respectively.) The data on P. M. Everett and his descendants were evidently supplied by his eldest surviving daughter, Emma Eugenia Everett Ball. Her autographed copy, dated 1904, contains a few supplementary notes concerning a son and a brother. Surprisingly, it states Everett's birth year as 1812, instead of 1807. But 1812 also appears with "10-21-1812 ?" on cemetary records. A birth date in 1812 is inconsistent with birth at Winchester.
In 1799, a group from Winchester, including Seth Hills, Philo Everett's maternal great grandfather, founded Vernon in Oneida County, New York. Andrew Everett (Philo's grandfather) and Statira Hills Marshall Everett moved there in 1809. Philo's parents (Elihu and Roxy Marshall Everett) are found in Winchester until 1809, but were named in an 1812 deed as being "of Vernon". Roxy was a child of Statira's first marriage and it is probable that Elihu and Roxy moved when their parents did, in 1809.
See S. Boyd, <u>Annals of Winchester, Conn.</u> (Hartford,1873). For copy, see NEHGS=C. (The author has a copy, autographed "P. M. Everett, Marquette, 1875" and by Mrs. Ball.) It contains the above information on the 1812 deed of Elihu and Roxy Everett, but nothing on their children.

See also an extensive article by Charles Raymond Everett (a grandson), "The Philo M. Everett Family", Vol. 22, <u>Michigan History Magazine,</u> pp.430-442, Autumn Number 1938, hereinafter "CRE".

See F. B. Stone, <u>Stone, Ball, Everett and Johnson Ancestry</u> (Summit, NJ, 1990, copy NEHGS=C) and other sources cited therein.

6  Even when his sight was failing in 1886, Everett said: "It is no
hardship for me to write and in fact it is all I can do". He wrote a
clear hand on ruled paper even when he could not read what he had
written.

7  For short histories of the Canal, see Andrist, Erie Canal in
The American Heritage Junior Library (New York, 1964) and S. H. Adams,
The Erie Canal (New York, 1953). A more detailed study is Shaw, Erie
Water West, a History of the Erie Canal, 1792-1854, (University Press
of Kentucky, 1966). For a map, Canals of New York State, see
A Canalboat Primer, p.4 (1981, The Canal Museum, Syracuse, N.Y.)

8  See Krause, The Making of a Mining District, Keweenaw Native Copper,
1500-1870, Wayne University Press (1992), hereinafter "Kr". Although,
as indicated, the area covered is confined to the Copper Country of
Michigan, it contains a comprehensive analysis of the three great
personalites (Burt and Jackson, as well as Houghton) who did so much to
shape this narrative.

9  Charles T. Jackson, like Houghton, was first trained as a physician. Of
an old Massachusetts family, after graduation from Harvard Medical
School, he travelled in Europe and became interested in geology. This
led to a position as geologist for three Eastern states. Kr pp. 182-
183, 199-201.

10  Kr. pp. 182-183, 199-201

11  Several years later, Jacob Houghton, one of Burt's party, made much of
the excitement and the gathering of specimens. (His "much more
expansive" reconstruction is quoted by Hatcher in his authorized
biography of the Cleveland Cliffs Company, A Century of Iron and Men,
Bobbs-Merrill (1950), pp.24-25), hereinafter "Hat".)

12  See Geological Survey of Michigan Upper Peninsula 1869-1873 by Board of
Geological Survey (N.Y., Julius Bien 1873), hereinafter "Geol. Surv."
for a "Historical Sketch of Discovery and Development", esp. pp.11-19.
It includes a map (p.14), supposedly made in 1844 by Wm. Ives that
shows "Iron hills" west of this area. However, it also shows the
correct location of Teal Lake, which was not shown on the Houghton-Burt
maps of 1845-1846 and was not known until 1846. See below.
  As Krause points out (Kr. p.176 et seq.), efforts were made promptly
after Dr. Houghton's death to organize the results of the combined
linear and geological surveys in the Upper Peninsula, in which "William
Burt and Bela Hubbard had been Houghton's primary assistants in that
fieldwork" They were asked to assemble reports, which were published

in 1846 by Jacob Houghton, Dr. Houghton's brother. His report, <u>The Mineral Region of Lake Superior</u>, (Buffalo,1846) contains Burt's report (pp.84 <u>et seq.</u>) and Hubbard's (pp.104 <u>et seq.</u>), as well as reprints from Dr. Houghton's work. Very interesting maps were attached, giving locations made up to July 17, 1846. (See detail in Plate11, discussed below.)

  One can look in vain for any report of Burt's "discovery" of iron in 1844. This disposes of the statement that a report of Burt's, like that of Houghton, "generated a fever of speculation and dreams of wealth among men in far places" or that "In his store in Jackson, Mich., Philo M.Everett read these reports, heard about them from travellers and mulled the possibilites" as stated in <u>A Bond of Interest</u>, a reprint of <u>Harlow's Wooden Man</u>, MCHS (Fall 1978), apparently sponsored by The Cleveland-Cliffs Iron Company.

[13] See Ex. Doc.1 - 31st Cong., 1st Session, <u>Message from the President, Dec 24, 1849, Part III</u>, which includes the report (hereinafter "Jack") of Professor Charles T. Jackson, U.S. Geological Surveyor of the Mineral Lands of the United States in Michigan

[14] These reports, that mentioned in the text as well as one by a Mr Barbeau that seems based on the Jackson party, were described in some detail, followed by Everett's report, and the conclusion:

        That Mr. Everett was really the pioneer in the discovery and development of the Lake Superior Iron mines, cannot be successfully disputed. It is true others may have visited the Jackson mountain about the same time, but we have no evidence that any of them discovered or knew of the existence of its hidden treasures.

[15] Philo M. Everett, <u>Recollections of the Early Explorations and Discovery of Iron Ore on Lake Superior,</u> Vol.11 of Michigan Pioneer Collections (1887) at pp. 161 to 174 (hereinafter "Ev.Rec.") The first part of these recollections is contained in Havighurst, <u>The Great Lakes Reader,</u> (Collier Books) (1966), with a somewhat inconsistent introduction.

The author's typescript of these recollections contains this attachment:

    "Jottings of the first discovery of iron at Marquette and the incidents attending it from the spring of 1845 until after the companies began shipping in 1856. It is forty-one years since the commencement of explorations, and my only object is to get the dates and the incidents together, so as to form a readable article.
                          February 1887
                          P.M.Everett"

[16] See Kr., cited above

<sup></sup>17 Record in <u>Compo v. The Jackson Iron Company</u>, 49 Mich. 39-75, 12 NW 901-903 (1882); 50 Mich. 578-596, 16 NW 295-303 (1883); 76 Mich.298-510;43 NW 602-606 (1889), hereinafter "Transcript" or "Tr". As well as in the official archives, a copy of the transcript is in the John M. Longyear Research Library of the Marquette County Historical Society (MCHS) from which many quotations have been taken.

18 Ev. Rec.161 et seq.

19 S. T. Carr, W. Monroe and Rockwell are named in the Transcript.

20. Ev. Rec. 162

21 This clearly indicates Everett's intention to proceed promptly to the Copper Country to pursue the search for copper.

22 A good description of the shipping in 1845 is found in the reminiscences of "one of the survivors", Lewis Marvill, reprinted as "Voyage of the Independence" at p. 259 of <u>The Great Lakes Reader</u>, cited above. See also Williams, <u>The Honorable Peter White</u> (Penton Publishing Co., Cleveland, 1907), Chapter XII, pp.109-117.

23 Samuel Ashman was a long-time employee of the American Fur Company. He had met Governor Cass's party at the headwaters of the Mississippi in 1820 and was a prominent citizen of Sault Ste. Marie from 1824 onward, assisting travellers, such as Everett and Douglass Houghton's party

24 $50 from each of the 13 members.

25 In 1945, Ray A. Brotherton of Negaunee prepared an article and slide show. It was presented in Negaunee and at the annual meeting of the Marquette County Historical Society and the article was printed in <u>The Daily Mining Journal</u>, Marquette, Mich Oct. 15 & 17, 1945. (He said that much of his material was given to him in talks with Henry Van Dyke and from data supplied by C. R. Everett.) It is a very florid article and begins with an impossible tale of how the Everett party started out by rail and by stagecoach to Grand Rapids, whereupon they hired a "tote wagon" and drove up the Lower Peninsula of Michigan, over 200 miles, without a road or bridges, in four days ! This is contrary, of course, to Everett's account of 1887 (Ev. Rec. p.162), quoted within. Nevertheless, I have retained some of Brotherton's descriptions when they seemed consistent with the record. Brotherton wrote other accounts of the Jackson Mine, some of which included other fanciful bits and I have not considered it appropriate to include or take issue with them.

[26] Jack p.477.

[27] Ev. Rec.161 et seq.

[28] Louis Nolan, who is one of the leading characters in this narrative, was another of the prominent citizens of Sault Ste Marie and was a former sheriff of that large county. His exciting history is contained in many anecdotes in Kidder, Ojibwa Narratives, Bourgeois ed. (Wayne State Univeristy Press, 1994), hereinafter "Kidder". These were related to Kidder by Kawbawgam and Nolan's son, Francis, aka Jacques LePique. (Cf.p.15 and pp.129 et seq.)

As the principal doorway to the Lake Superior Country, the Sault had developed a population with a strong mixture of French and Indian ancestry, especially such famous persons as John Johnston, his Indian princess and their family, which included their son-in-law Henry Schoolcraft, the great historian of the Indians.

[29] Ev. Rec.163 et seq.

[30] Ev. Rec.165

[31] The decisions are cited as Compo v.Jackson Iron Co. 49 Mich 39, 12 NW 901 (1882) and 50 Mich 578, 16 NW 295 (1883) and Kobogum v.Jackson Iron Co. 76 Mich 498, 43 NW 602 (1889). The transcript on appeal (the Transcript) was used by Justice John Voelker, pen name Robert Traver, in his highly fictionalized novel, Laughing Whitefish. (See fn. 2 above.) Both publicly and to members of the family, he recognized that his story sought "to create illusion", rather than to deal "with sober fact".

[32] 43 NW at 602

[33] 16 NW 295 at 298

[34] For Everett's testimony, see Tr.38 to 68

[35] Geol. Surv. pp. 14-15.

[36] Ev. Rec. pp.164-165 fails to recall that Nolan could not find the "ore" that he had described at the Sault and that it was necessary to send to find Marji Gesick to show the "iron hills".

[37] See Souvenir Program, Marquette County Centennial Pageant, 1836-1936, which, after reporting how Burt found the presence of iron ore in 1844, casts doubt on accounts of mineral discoveries as follows (unnumbered page):

"Later on fictitious accounts arose, involving rooting pigs [an allusion to the legend of Calumet & Hecla] and fallen trees. There were, of course, no pigs to root at this early day, and no tree was large enough to hide the great protuberences of iron ore that stood out high above the surface of the land, some of it glinting in the sunlight like burnished steel as P. M. Everett described it."

38 While the "fallen pine tree" legend is now generally debunked, it still endures, cast in bronze. A monument was erected in 1904 by the moribund Jackson Iron Company near the original mine, but it was vandalized and moved to a park in Negaunee. The bronze plaque on it states that the first discovery of iron ore "was found under the roots of a fallen pine tree, in June 1845, by Marji Gesick . . ". See back cover of Michigan History for November/December 1994. The plaque is incorrect in several respects.

39 Frank Matthews (1902-about 1983) for many years maintained "The Jackson Mine Museum" on U.S. 41, just east of Negaunee, crammed with pictures and artifacts of the Mine. See Lyman, The Mirrored Wall, (1973, Globe Printing, Ishpeming, MI). Much of this treasure is now in the Michigan Iron Industry Museum.

40 Ev. Rec. 165

41 As later testified, this was the permit in the name of James Ganson. It is set out in full in the Transcript (p. 114 et seq.), followed by the formal selection of the tract, executed at Fort Wilkins October 4, 1845 by E.S. Rosknell (sic), Att'y and by the certification of Lease 133 as prayed for.

42 See Kr. pp. 176-177 for the circumstances surrouding this and subsequent reports.

43 This was the Report of A. B. Gray, dated March 10, 1846 (Doc. No. 211, H. R. 29th Cong. 1st Session), hereinafter "Gray".

44 Tr. p. 123

45 This characterization has been attributed to several sources. (See Sibley's remark, infra.)

46 Ev. Rec. 165

47 These were, of course, in pursuance of their principal objective -- copper. Jacob Houghton's List of Locations (op. cit.)pp. 172 et seq. has the following permits:

593-James Ganson (Lease 133, date Dec 12, 1845), 625-P M Everett; 633-E S Rockwell (after 7/17/45), 825-F Farrand, 881-J W Carr, and 882-S T Carr It also reports that 1/2 of the Everett and Rockwell permits were lost as conflicting with 370 None of these permits (other than Ganson's) was found on his or Gray's maps (See Plates 2,11 and 12 )

48 Ev Rec pp 166 et seq

49 This is part of the "Route followed in September 1845" shown on the Gray Map (Detail Plate 13), that Gray used, of which he says, " at the head of the lake [Vieux Desert], we struck upon an old Indian trail; and following it through swamps and ravines, over hills and high granite ranges, we reached the Anse settlement on the fourth day The trail was a very rough one, and if it had not been for the quick eye of one of our Indian voyageurs, we should have found it extremely difficult to make our way through" Gray, p 11

50 See Gray, pp 6 et seq Gray had also been at the treaty payment and, as indicated in the previous note, was returning to the lake up the same trail

51 These dates and this sequence are hard to reconcile with the fact that the claim at Teal Lake was "staked out" "about the 20th day of September" (Tr 63) and permits of W H Monroe and P M Everett were "laid" with Gen Stockton at Copper Harbor on "Oct 6" See Fadner, Fort Wilkins and the U S Mineral Land Agency, (New York, 1966) p 235 If, as stated above, Everett left from Copper Harbor on this trip, it must have been October That would give scant time to complete it and get to Jackson by October 24, as stated in the subsequent material

52 Gray called the Indians on the route "the wildest Indians" (and they were the ones so described by Everett) At Lake Vieux Desert, however, they grew some very fine potatoes on several acres of cultivated ground and "This is what gives to the lake the name of the 'Old Gardens,' or 'old planting grounds'" Gray p 10-11

53 Gray describes a violent equinoctial gale on September 19, 1845 and how "the universally lamented Dr Houghton" was "lost on the 13th October" Gray p 11

54 This was the last place where Everett worked before moving to Jackson in 1840 Gilbert Johnson worked with him and continued there Now Everett exhibits plans of future operations Gilbert moved to Lake Superior about 1859 and became Superintendant of the large Lake Superior Mine

55 Ev. Rec. 167-8

56 Clearly he had not heard of Burt's township lines earlier. See Transcript, quoted above.

57 Ev. Rec. 168

58 A precise southwest course would make it just over 0.707 miles.

59 This correction, made in the summer of 1846, does not appear in the 1846 map published by Jacob Houghton, Jr. earlier that summer (Plate 11) nor in the Gray map, originally submitted with his report on June 16, 1846 and published in a revised version on January 9th. 1847. (See Plates 2 and 12.)

60 Letter of Oct. 21st, 1870, quoted in part in Geol. Surv. pp.15 et seq

61 This transaction is reported in the history of the Cleveland Cliffs Company, Hat. pp. 47 et seq.

62 See 43 NW 602 at 603

63 Tr. p. 59

64 Ev. Rec. 168

65 Ev. Rec. 168

66 This is the dwelling from which the Clark-Graveraet-Harlow group briefly evicted them (as narrated subsequently). Micklow emphasizes that it was the first white residence in Marquette. Micklow p. 17.

67 From July 24 to July 26, 1847, under instructions from Professor Jackson, Dr. John Locke, assistant geologist, visited Mr. McNair at the dam building on the Carp River (T.48, R.26W) nine miles from the mouth of the river. He also reported on the Jackson iron mine (T.47, R.27), which "will be in section 1, when surveyed." He extolled the quantity and quality of the ore. (Jack. pp. 574-575).

68 Tr p. 110

69 See the yearend report made by F. Farrand, President, Jackson Mining Co., of Jackson. Tr. pp. 188-191.

70 Forge, p. 8. This source is a very complete and accurate report and is relied upon heavily, sometimes without attribution

[71] Ernest H. Rankin, "From a Forge a Town was Born", Vol 20, No 1, Inland Seas, Spring 1964, pp.33-34, hereinafter Rankin. (In files of MCHS.) Ernest H. Rankin for many years collected reminiscences and memorabilia from many sources (including Everett's descendants) for the Marquette County Historical Society, of which he was Executive Director until he retired in 1969
He inspired others to pursue the history of the Iron Industry and even in retirement was still searching for means to honor Charlie Kawbawgam and Marquette's "pioneer citizen, Philo Marshall Everett ...the man who opened the Marquette Iron Renge". If his writings contain inaccuracies, they should be attributed to faulty memories of elderly residents whom he interviewed. The subseqent studies on Kawbawgam's and Everett's lives and the Carp River Forge owe much to the initiative that he gave to Marquette history.

[72] Ev. Rec. 168

[73] Her "relation", contained in the Daily Mining Journal of Nov. 2, 1922 and alluded to in her obituary of June 3, 1923, maintained that this was in November 1847 and that as "a ten-year-old girl",she was the first white child in Marquette. However, she was not ten until 1848 and it is unlikely that she would have come so late in the year, as a girl would not have wintered in the area at that time. This was one of several "firsts" that leave the listener in a questioning state. Her cousin Emma Everett (Ball) confirmed that she came in 1848

[74] Rankin, cited above.

[75] Ev. Rec. 168. See also Carp River Forge; A Report. (cited above in the text and herein as "Forge"), which provides both a historical and an archeological reconstruction of the forge, which it characterizes as "a crude and primitive establishment." (p. 26)

[76] Ev. Rec. p. 169 et seq.

[77] Following a lengthy controversy surrounding Professor Jackson, Foster and Whitney were instructed to take over . They produced a respected report, See Foster & Whitney, Report on the Geology of the Lake Superior Land District, Part II, The Iron Region, Executive No.4,(Senate, March,1851), herein "F&W". See also Kr. 187 et seq.

[78] Tr.193 et seq. and The Chronicle, (Marquette, June 5, 1909).

[79] See also J. S. Burt, They Left Their Mark, a Biography of William Austin Burt. (Landmark Enterprises, 1985), hereinafter "Burt", p.107.

[80] Micklow p.36 and Marquette Mining Journal, Jan.20,1872

[81] Ev. Rec. 171

[82] This apparently consisted of 200 shares. Their value might have been anywhere between $12 and $50 per share, as they varied radically. See Forge and Micklow.

[83] CRE pp.435-436

[84] Her letter is a part of family papers deposited in the Michigan Iron Industry Museum.

[85] Daily Mining Journal, Feb. 6, 1888. See also ibid. Feb. 1, 1888.

[86] Besides the Burt family, these were Mr. and Mrs. Charles Johnson with their niece (adopted daughter), Mary Huntoon, and the Everett family -- parents and four children, Emma Eugenia, Edward Philo, Charles Marshall and Mehitabel Ellen (Aunt Ella) and finally a baby, Katherine Eliza Johnson Everett (Kitty).

[87] CRE pp. 438-441

[88] This was apparently not Charles Kawbawgum, who married Margi Gesick's daughter, Charlotte, and who also became a part of this narrative. For his history, see Kidder, pp. 14-17 and passim.

[89] The Chief died about 1862, near Presque Isle, in the boat of Louis Nolan's son, Francis (aka Jack or Jacques LePique), who had gone to pick him up in his last illness. He was buried in an Indian cemetary near the Carp River Forge, but the spot is now unknown, kept secret, Frank Matthews told me, for fear of vandals. (See Rydholm, Superior Heartland, a Backwoods History [of Marquette], (Published Privately, 1989), hereinafter "Rydholm", at p.139.

[90] Ev. Rec. 171-172

91 A gift of spectacles and the letter from Mr. and Mrs. Fraser were delivered at the golden wedding anniversary, hereinafter described, and was included in the newspaper account and is reproduced here, rather than in the text:

Marquette, November 19th 1883.

Mr. and Mrs. P. M. Everett:

Dear and Valued Friends: -- Mrs. Fraser and myself cannot let this opportunity pass without saying a few words on this, your fiftieth anniversary or "Golden Wedding," and would ask you accept the small token of our high esteem that accompanies this letter. How well do I remember when I was a boy, (in 1855), a stranger among strangers here, how you took the opportunity to give me good counsel and advice, furnished me good books to read, and how you opened your house, not only to me but to other young men, and made it pleasant for us. and I know I never have forgotten, nor shall I ever forget, your kindness as long as I live. I feel that the young and old of Marquette owe you a debt of gratitude for the social and moral influence you have exerted over this, your adopted city, which can never be repaid; and the prayer of your humble servant is that you may long be spared to us, and that your declining years may be as bright as these "golden eyes." How many of us that still remain in this county, and many more that have gone, can say as Christ said, "I was a stranger and ye took me in;" "I was athirst and ye gave me to drink;" "I was an hungered and ye fed me;" "I was sick and ye visited me." This and much more can be said of you, and the only regret we have is that your days are "gliding swiftly by", as with the rest of us. That peace and happiness may be and remain with you is the wish of

Yours sincerely,
Mr. and Mrs. Edward Fraser

92 Ev. Rec. 171-172

162

[93] See Forge, pp.15-16 for a description of the Eaton and Curtis Managements.

[94] Rankin p. 40

[95] Ev. Rec. 172-174

[96] Printed in an article entitled "IN YE OLDEN TIME" in the Marquette Mining Journal. January 1, 1897.

[97] Ev. Rec. 174

[98] As noted above, there are different spellings throughout of "Marji Gesick" and "Charlie Kobogum", some authors pointing out (quite unnecessarily) that this is because the Chippewas (or Ojibwas) did not have a written language.

[99] Forge (p.33) points out that the decision includes "facts not available elsewhere", but the record contained in the Transcript is much much more helpful.

[100] This issue was treated at length in "Property Rights and Indian Marriage", Vol. XL Michigan Alumnus Quarterly Review, pp.154 et seq. (July, 1934).

[101] Brinks, Peter White: a Career of Business and Politics in an Industrial Frontier Community, a Ph.D. dissertation at the University of Michigan (1965), hereinafter "Brinks", copy in MCHS. This, together with a somewhat popularized biography by Brinks, Peter White (Eerdmans, Grand Rapids, 1970), hereinafter "Brinks PW", serve as antidotes to the life-long publicity campaign that White elicited and which seems to have culminated in the adulatory biography above cited, The Honorable Peter White. See also commentary on the White biographers in Brinks PW pp.50-51. But, even more, see a collection of heightened legends and myths, "The First Man of Marquette, Sketches of Hon. Peter White" from the Marine Review, Cleveland, Ohio, printed in Vol.30, Michigan Pioneer & Historical Collections, pp.119-140 (1905).

[102] See the attack in a 1853 letter quoted by Brinks (p.25) and by Micklow

[103] Rydholm pp.315-316

104 Statistics show that at this period, the dollar bought some 2.5 times as much as in 1967, but that reflected the aftermath of the Civil War. For most of the hundred years before World War I, the multiple was closer to 4.

105 The quotations in this section are from his unpublished memoir in the family papers, unless otherwise indicated.

106 See _Plaindealer--Extra_, Marquette, L.S. June 15, 1868. (Copy in MCHS)

107 Letter in family papers.

108 White quoted in Brinks, pp. 139-140

109 Holbrook, _The Age of Moguls_, (New York, 1953) pp. 55-56

110 Brinks PW p. 31

111 Brinks also points out that, although White's banking career was not without blemish (Brinks PW), in his later years he became a major philanthropist, leaving remarkable contributions to the city and donating generously to a church and the library. Largely through his gifts and efforts, the city has beautiful parks, above all those on and surrounding Presque Isle.

112 CRE p. 437